Proceedings
of the Steklov Institute of Mathematics

edited by
S. M. Nikol'skiĭ

Number 121 (1972)

Logical and Logico-Mathematical Calculi. 2

edited by
V. P. Orevkov

AMERICAN MATHEMATICAL SOCIETY
Providence, Rhode Island
1974

Академия Наук
Союза Советских Социалистических Республик

ТРУДЫ
ордена Ленина
МАТЕМАТИЧЕСКОГО ИНСТИТУТА
имени В. А. СТЕКЛОВА
CXXI

Ответственный редактор
член-корреспондент АН СССР *С. М. НИКОЛЬСКИЙ*

ЛОГИЧЕСКИЕ И ЛОГИКО-МАТЕМАТИЧЕСКИЕ ИСЧИСЛЕНИЯ · 2

Сборник работ
под редакцией В. П. ОРЕВКОВА

Издательство „Наука"
Ленинградское отделение
Ленинград 1972

AMS (MOS) subject classifications (1970).
Primary 02B10, 02C15, 02G05, 02G99, 68A40; Secondary 02B10, 68A40.

Library of Congress Cataloging in Publication Data CIP

Logical and logico-mathematical calculi. 2.

(Proceedings of the Steklov Institute of Mathematics, no. 121 (1972))
Translation of Logicheskie i logiko-matematicheskie ischislenii͡a. 2.
Called "2" in continuation of The Calculi of symbolic logic, 1, which was published in 1971.
1. Logic, Symbolic and mathematical. I. Overkov, V. P., ed. II. Series: Akademii͡a nauk SSSR. Matematicheskiĭ institut. Proceedings, no. 121, 1972.
QA1.A413 no. 121 [QA9] 510'.8s [511'.3] 74-8854
ISBN 0-8218-3021-X

PROCEEDINGS OF THE STEKLOV INSTITUTE OF MATHEMATICS
IN THE ACADEMY OF SCIENCES OF THE USSR

(Труды математического института им. В. А. Стеклова, т. CXXI, 1972)

TABLE OF CONTENTS

EDITOR'S PREFACE

This volume is the second collection of papers from the seminar on mathematical logic at the Leningrad Branch of the Steklov Mathematical Institute of the Academy of Sciences of the USSR. The collection consists of papers on the theory of logical inference and its application to the construction of algorithms for machine search for inference. Increasing recent interest in this domain of mathematical logic is connected with the development of mathematical cybernetics, in particular those of its divisions in which the question of simulation on computers of complex forms of human intellectual activity is considered.

In the papers of this collection calculi of constructive logic are considered as well as calculi of classical logic. But here calculi of the second type never occur as the logical basis of any investigations; they are considered only as constructively defined objects of study and, like constructively defined objects of other types considered in this collection, they are investigated on the basis of principles of the constructive trend in mathematics.(1) In this sense it is possible to say that this collection belongs to the constructive trend in mathematics and borders on the series *Problems of the constructive trend in mathematics.* (2)

All the papers published in this collection were completed and delivered at seminars in 1965–1971.

V. P. Orevkov

---

(1)The main lines of the constructive trend in mathematics are formulated, for example, in the paper by A. A. Markov, *On constructive mathematics* (Trudy Mat. Inst. Steklov. 67 (1962), 8–14; English transl., Amer. Math. Soc. Transl. (2) 98 (1971), 1–9).

(2)See Trudy Mat. Inst. Steklov. 52 (1958); 67 (1962); 72 (1964); 93 (1967); 113 (1970) [for English translations of most of these papers, see Amer. Math. Soc. Transl. (2) 23 (1963); 29 (1963); 57 (1966); 64 (1967); 94 (1970); 97 (1971); 98 (1971); 99 (1972); 100 (1972); *Constructive real numbers and constructive function spaces*, Amer. Math. Soc., Providence, R. I., 1968; Proc. Steklov Inst. Math. 93 (1967); 113 (1970)].

Trudy Mat. Inst. Steklov.
121 (1972)

Proc. Steklov Inst. Math.
121 (1972)

ON A BOUND FOR THE COMPLEXITY OF TERMS IN THE RESOLUTION METHOD

UDC 51.01 : 164

N. K. ZAMOV

ABSTRACT. A term $f^{(n)}t_1\cdots t_n$ is called regular if any of its subterms having positive rank has the form $g^{(m)}t_1\cdots t_m$, where $m < n$. A number of lemmas on regular terms are proved. In particular, it is shown that the rank of a regular term $f^{(n)}t_1\cdots t_n$ is not greater than $n - m + 1$, where n is the degree of the functional symbol $f^{(n)}$ and m is the minimal degree of the functional symbols occurring in $t_1, \cdots, t_n$. In this article some classes of formulas are indicated for which, when provability by the resolution method has been established, it is sufficient to construct derivations containing only disjunctions with regular terms. In particular, for the class of almost–Skolem formulas one need construct only derivations such that the ranks of the terms in the disjunction do not exceed 1.

Bibliography: 5 items.

The resolution method, described in [1], is, together with the inverse method (see [2]), one of the most widely used methods of establishing provability of formulas of the predicate calculus.

In the process of establishing provability of formulas by the resolution method(1) one constructs a certain set of initial disjunctions with respect to each formula. Provability of a formula is deemed to be established if from the set of initial disjunctions we can derive a disjunction of a special form (the empty disjunction) according to a certain rule (rule P). The complexity of the objects derived in this way (the number of elementary formulas in the disjunction and the ranks of the terms(2) appearing in them) can grow without bound.

In this note we call attention to certain classes of formulas for whose establishment by the resolution method one need use only those disjunctions in which the ranks of the terms do not exceed some preassigned number depending only on the form of the initial formulas. In particular, for formulas of the type

$$\exists x_1 \ldots \exists x_n \forall y_1 \ldots \forall y_m \mathfrak{A}(x_1, \ldots, x_n, y_1, \ldots, y_m),$$

AMS (MOS) subject classifications (1970). Primary 02B10, 02G99; Secondary 68A40.

(1) In the present note the resolution method will be formulated in terms of provability rather then realizability as in [1].

(2) The definition of the rank of a term is given below.

where $\mathfrak{A}$ is a formula without quantifiers, it is sufficient in establishing provability by the resolution method to use only disjunctions whose terms have rank not greater than 1.

We shall introduce several definitions needed in the exposition to follow, most of which are found in [1].

The following symbols are *variables*: x, y, z, x_1, y_1, z_1, etc.

For each $n \geqslant 0$ the following are *n-place functional symbols*: $g^{(n)}$, $h^{(n)}$, $g_1^{(n)}$, $h_1^{(n)}$, etc.

For each $n \geqslant 0$ the following are *n-place predicate symbols*: $Q^{(n)}$, $R^{(n)}$, $Q_1^{(n)}$, $R_1^{(n)}$, etc.

We next define the concepts of *term* and *rank of a term.*

a) Variables and 0-place functional symbols are terms of rank 0.

b) If $t_1, \cdots, t_n$ are terms and f is an n-place functional symbol, then $ft_1 \cdots t_n$ is a term whose rank is greater by 1 than the maximal rank of the terms $t_1, \cdots, t_n$.

We denote the rank of the term t by $r(t)$.

We define the concept of a *subterm* of a term t:

a) The term t is a subterm of itself.

b) If t has the form $ft_1 \cdots t_n$, then all subterms of the terms t_i $(i = 1, \cdots, n)$ are subterms of t.

A string of symbols consisting of an n-place predicate symbol followed by n terms $(n \geqslant 0)$ is called an *atomic formula.* An atomic formula or its denial will be called a *literal.* A pair of literals one of which is the denial of the other is called a *contrary pair.* A finite (possibly empty) set of literals is called a *disjunction.* The empty disjunction is denoted by $\square$.

An expression of the form t/v, where v is a variable and t is a term not identical with v, is called a substitution component. A finite (possibly empty) set $\{t_1/v_1, \cdots, t_n/v_n\}$ $(n \geqslant 0)$ of substitution components, where $v_1, \cdots, v_n$ are pairwise distinct variables, is called a *substitution.*

Let E be an arbitrary finite sequence of symbols (functional symbols, predicate symbols and variables) and σ a substitution of the form $\{t_1/v_1, \cdots, t_n/v_n\}$. An operation consisting of the replacement of all occurrences of the variable v_i in E by an occurrence of the term t_i $(i = 1, \cdots, n)$ is called a *concretization* of E by means of σ. The symbol sequence thus obtained is denoted by $E\sigma$ and is called a *σ-instance* of E.

Let C be a set of finite sequences of symbols and σ a substitution. The set of all finite sequences of symbols of the form $E\sigma$, where E occurs in C, is called a *σ-instance* for C. We shall denote this set by $C\sigma$.

Let C be a finite set (or a finite set of finite sets) of symbol sequences, and let $v_1, \cdots, v_n$ $(n \geqslant 0)$ be all the variables occurring in these sequences. Let us call the substitution $\{x_1/v_1, \cdots, x_n/v_n\}$ an *x-standardization* of the set C and $\{y_1/v_1, \cdots, y_n/v_n\}$ a *y-standardization.*

By the *composition* of the substitutions σ and λ, denoted by $\sigma\lambda$, we mean a substitution μ such that $E\mu = (E\sigma)\lambda$ for any finite sequence (or set of finite sequences) of symbols of E.

A substitution σ is called a substitution *unifying* the finite set E of symbol sequences if $E\sigma$ is a set with a single member. A substitution σ unifying the set E is called *general-unifying* for E if for every substitution λ unifying E a substitution θ exists such that $\lambda = \sigma\theta$.

We will say that for the disjunctions $N, C, D,$ the substitutions ξ, η, σ and the sets of literals L and M the relation $P^*(N, C, D, L, M, \xi, \eta, \sigma)$ is satisfied if the following conditions hold:

a) $L \subseteq C$ and $M \subseteq D$.

b) The substitutions ξ and η are, respectively, the x-standardization of the disjunction C and the y-standardization of the disjunction D.

c) The substitution σ is general-unifying for the set of all atomic formulas occurring in $L\xi \cup M\eta$.

d) $L\xi\sigma$ and $M\eta\sigma$ are singleton sets whose elements constitute a contrary pair.

e) N is the set $(C - L)\xi\sigma \cup (D - M)\eta\sigma$.

If $P^*(N, C, D, L, M, \xi, \eta, \sigma)$ is satisfied for some $N, C, D, L, M, \xi, \eta, \sigma,$ then we will say that the disjunction N is obtained from the disjunctions C and D by the rule P. We will call formulas occurring in L and M *principal formulas,* while C and D will be called the *premises* of the application of the rule P.

By a P-*derivation* of the disjunction D in the set of disjunctions S we will mean a list of disjunctions each of which either belongs to S or is obtained from the preceding disjunctions in the list by rule P,(3) and the last disjunction of the list is D.

Suppose that the closed formula $\mathfrak{A}$ has the form

$$\pi_1 u_1 \ldots \pi_n u_n \bigvee_{i=1}^{m} \mathop{\&}_{j=1}^{\delta_i} A_{ij} \, (n \geqslant 0, \ m > 0, \ \delta_i > 0),$$

where A_{ij} is a literal $(1 \leqslant i \leqslant m, \ 1 \leqslant j \leqslant \delta_i)$ and $\pi_e u_e$ is a quantifier complex $(1 \leqslant e \leqslant n)$, and let $v_1, \cdots, v_r$ $(1 \leqslant r \leqslant n)$ be a complete list of variables of formula $\mathfrak{A}$ connected by universal quantifiers and enumerated in the order of their appearance in the prefix of formula $\mathfrak{A}$. Denote by A_{ij}^* the literal obtained from A_{ij} $(1 \leqslant i \leqslant m, \ 1 \leqslant j \leqslant \delta_i)$ by replacing each occurrence of the variable v_q

(3) We shall assume that every P-derivation in S is provided with an analysis of the derivation, i. e. with an indication for each disjunction of the derivation as to whether the disjunction belongs to the set S or from which disjunction it is obtained by rule P, and information about the kinds of principal formulas, standardizations and general-unifying substitutions of this application of rule P.

$(1 \leqslant q \leqslant r)$ by the term $f_q^{(p)} w_1 \cdots w_p$, where $w_1, \cdots, w_p$ are complete lists of variables connected by existential quantifiers preceding the quantifier complex $\forall v_q$ in the prefix of formula $\mathfrak{A}$, and the variables $w_1, \cdots, w_p$ appear in the order of their occurrence in the prefix of formula $\mathfrak{A}$.

The set of all disjunctions of the form

$$\{A^*_{i,1}, \ldots, A^*_{i,s_i}\} \quad (i = 1, \ldots, m)$$

is denoted by $S_{\mathfrak{A}}$.

In [1] a theorem was proved (resolution theorem), which in our context has the following form: for formula $\mathfrak{A}$ to be proved, it is necessary and sufficient that there exist a P-derivation of the disjunction $\square$ in the set $S_{\mathfrak{A}}$.

For an arbitrary formula $\mathfrak{A}$ it is obvious that there can exist a number of **P**-derivations of the disjunction $\square$ in $S_{\mathfrak{A}}$. In some cases it turns out to be convenient to use so-called complete tactics to look for P-derivations. When a complete tactic is used, one does not in general construct all P-derivations possible but only those having the specific properties characterizing that tactic. Complete tactics guarantee the construction with these properties of the disjunction $\square$ in $S_{\mathfrak{A}}$, if the disjunction $\square$ is deducible in $S_{\mathfrak{A}}$.

For certain classes of formulas we can determine complete tactics such that in disjunctions derived by their application the number of literals or the ranks of the terms or both together will not exceed some specified number (in the last case the class of formulas turns out to be solvable).

We will describe below classes of formulas and determine tactics whose application necessarily leads to disjunctions wherein the terms have a special form that allows one to assert that the ranks of these terms cannot be larger than some specified number.

In order to formulate the tactics and to describe the kind of terms occurring in the P-derivations constructed thereby, we introduce a few more definitions. It will be convenient below to represent the P-derivations in the form of trees.

Suppose we are given a P-derivation of a certain disjunction N in S. By a *derivation tree* of the disjunction N (or simply a tree for N) we shall mean a configuration constructed as follows:

a) If $N \in S$, the tree for N is N itself.

b) If N is obtained by rule P from disjunctions C and D, the tree for N is a configuration of the form

where Δ_C and Δ_D are the trees for C and D, respectively.

The disjunct N and all the disjuncts occurring in Δ_C and Δ_D will be called the disjunctions occurring in the tree for N; N is called the *root* of the tree.

Let Δ be a tree and σ a substitution. By $\Delta\sigma$ we shall mean a tree obtained as a result of concretizing every disjunction occurring in Δ by means of the substitution σ.

Let Δ_N be a derivation tree of N in the set of disjunctions S. We define for Δ_N a corresponding propositional tree, which we denote by $\widetilde{\Delta}_N$:

a) If $N \in S$, $\widetilde{\Delta}_N$ is identical with Δ_N (that is, with N).

b) If for some $C, D, L, M, \xi, \eta, \sigma$ we have $\mathrm{P}^*(N, C, D, L, M, \xi, \eta, \sigma)$, then $\widetilde{\Delta}_N$ is a configuration of the form

$$\begin{array}{cc} \widetilde{\Delta}_C\tilde{\xi}\sigma & \widetilde{\Delta}_D\tilde{\eta}\sigma \\ \searrow & \swarrow \\ & N \end{array}$$

where $\tilde{\xi}$ and $\tilde{\eta}$ are the respective x- and y-standardizations of the disjunctions in $\widetilde{\Delta}_C$ and $\widetilde{\Delta}_D$, respectively, such that $L\tilde{\xi}$ is the same as $L\xi$ and $M\tilde{\eta}$ is the same as $M\eta$.

Evidently each appearance of a disjunction, literal or term in the tree $\widetilde{\Delta}_N$ is obtained by replacement of the variables by certain terms in the corresponding occurrence of the disjunction, literal or term in Δ_N.

If E is an occurrence of a disjunction, literal or term in the tree Δ_N, the corresponding occurrence in $\widetilde{\Delta}_N$ will be denoted by $\widetilde{E}$. By the *latent rank* of the occurrence of a term t in a tree Δ we shall mean the rank of the corresponding occurrence of the term $\tilde{t}$ in the tree $\widetilde{\Delta}$.

Let us divide the set of terms occurring in $\widetilde{\Delta}$ into classes such that the class k_{ij} contains terms of rank i beginning with j-place functional symbols $(i \geqslant 0, j \geqslant 0)$; moreover, we allow the class k_{00} to contain variables.

Let us fix some kind of partial ordering of the terms in $\widetilde{\Delta}$ in which each term of k_{ij} precedes an arbitrary term of k_{lm} if either $i > l$ or $i = l$ and $j > m$. Such an ordering of the terms we will call *regular*. Terms of the same class will be called *similar*.

Let E be a finite sequence of symbols or finite set of finite sequences of symbols. Terms that occur in E and for which there are no preceding terms in E are called *principal* in E.

Suppose that the literals E_1 and E_2 occur in a tree $\widetilde{\Delta}$. We will say that the literal E_1 *precedes* (*is similar to*) the literal E_2 if the principal term in E_1 precedes (is similar to) the principal term in E_2.

Let V_1 and V_2 be two different occurrences of terms in the tree Δ. We will say that V_1 *precedes* (*is similar to*) V_2 if $\widetilde{V}_1$ precedes (is similar to) $\widetilde{V}_2$. Let W_1 and W_2 be different occurrences of a literal in the tree Δ. We will say that W_1 *precedes* (*is similar to*) W_2 if $\widetilde{W}_1$ precedes (is similar to) $\widetilde{W}_2$.

The ordering of terms and literals in the tree Δ just introduced is called the *natural ordering* with respect to the regular ordering established above for terms in the tree $\widetilde{\Delta}$. It is clear that there can be different natural orderings of the same set of terms (literals), differing only in the method of ordering similar terms (literals).

In [4] it was proved that if a derivation tree Δ of the disjunct $\square$ in S exists, then for any fixed ordering of the literals in $\widetilde{\Delta}$ there also exists a derivation tree Δ_1 of the disjunction $\square$ in S in which the literals preceding all the other literals occurring in the premise are the principal formulas of the application of rule P. We call such a tree a tree with ordering of literals.

A tree with ordering of literals in which some kind of natural ordering with respect to the regular ordering of the terms is imposed is called a *tree with natural ordering of literals.* We will say that the term $f_1 t_1 \cdots t_n$ *covers* the term $f_2 s_1 \cdots s_m$ if $n \geqslant m$ and s_i coincides with t_i $(i = 1, \cdots, m)$. The term t is called *regular* if it covers all its subterms of positive rank.

LEMMA 1. *If the regular term $f_1 t_1 \cdots t_n$ covers the term $f_2 t_1 \cdots t_m$, then $f_2 t_1 \cdots t_m$ is a regular term.*

PROOF. Let $f_3 s_1 \cdots s_l$ be a subterm of $f_2 t_1 \cdots t_m$. If $f_3 s_1 \cdots s_l$ coincides with $f_2 t_1 \cdots t_m$, then $f_2 t_1 \cdots t_m$ covers $f_3 s_1 \cdots s_l$. Suppose that $f_3 s_1 \cdots s_l$ is different from $f_2 t_1 \cdots t_m$. Then $f_3 s_1 \cdots s_l$ is a subterm of one of the terms $t_1, \cdots, t_m$ and hence is a subterm of $f_1 t_1 \cdots t_n$. Since $f_1 t_1 \cdots t_n$ is regular, it covers $f_3 s_1 \cdots s_l$; therefore $l \leqslant n$, and s_i coincides with t_i $(i = 1, \cdots, l)$. Now $l < m$, since $f_3 t_1 \cdots t_l$ is a subterm of one of the terms $t_1, \cdots, t_m$ and cannot be a subterm of any one of the terms $t_1, \cdots, t_l$. From this it follows that $f_2 t_1 \cdots t_m$ covers $f_3 t_1 \cdots t_l$. Since $f_3 t_1 \cdots t_l$ is any subterm of $f_2 t_1 \cdots t_m$, we have proved that $f_2 t_1 \cdots t_m$ covers any of its subterms of positive rank, i.e. it is regular.

LEMMA 2. *Let v be a variable, u a term not containing v, and E a certain symbol sequence, not containing the variable v, such that Ev and Eu are initial subsequences of the symbol sequences that form regular terms t_1 and t_2, respectively. Under these conditions the term $t_1\{u/v\}$ is regular.*

PROOF. If the rank of the term u is 0, every subterm of positive rank of the term $t_1\{u/v\}$ is a $\{u/v\}$-instance of some subterm of t_1 having positive rank. Since concretization preserves the covering relation between terms and t_1 covers all its subterms of positive rank on account of regularity, $t_1\{u/v\}$ also covers all its subterms of positive rank. Thus $t_1\{u/v\}$ is regular.

Suppose that the rank of the term u is positive. Since u is a subterm of t_2, t_2 covers u. From this it follows that the number of symbols in E is no fewer than in u, but then t_1 also covers u. Hence every subterm of $t_1\{u/v\}$ of positive rank either is a $\{u/v\}$-instance of some subterm of t_1 of positive rank or coincides with u. The term u is covered by $t_1\{u/v\}$, since u is covered by t_1; but E does not contain the variable v; the other subterms of $t_1\{u/v\}$ of positive rank are covered by it because concretization preserves covering between terms. Therefore $t_1\{u/v\}$ is regular.

LEMMA 3. *If t_1 and t_2 are regular terms and σ is a general-unifying substitution for the set $\{t_1, t_2\}$, then the only term occurring in the set $\{t_1, t_2\}\sigma$ is regular.*

PROOF. It was shown in [1] that any general-unifying substitution for the set $\{t_1, t_2\}$ could be represented as a composition of single-term substitutions $\sigma_1, \cdots, \sigma_k$, where σ_i $(1 \leqslant i \leqslant k)$ is a general-unifying substitution for $\{t_{1,i}, t_{2,i}\}$, and $t_{j,i}$ for every i $(1 \leqslant i \leqslant k)$ is an initial subsequence of the symbol string constituting t_j $(j = 1, 2)$. Applying Lemma 2 k times, we find that the term t is regular.

A set of terms of positive rank is called regular if there is a term in it that covers all terms of the set. We note that this term is the principal term of the set for a natural ordering.

LEMMA 4. *If elimination of a term t from a set T of terms leaves a nonempty set, then this set is regular.*

PROOF. If t does not cover all terms of the set T, the lemma follows from the definition of a regular set of terms.

If t covers all the terms of T, then there is a term in $T-\{t\}$ which covers all the terms of $T-\{t\}$. Indeed, if t covers the terms $f_1t_1 \cdots t_n$ and $f_2s_1 \cdots s_m$ and $n \geqslant m$, then $f_1t_1 \cdots t_n$ covers $f_2s_1 \cdots s_m$. Therefore a term in $T-\{t\}$ beginning with the functional symbol of greatest degree(4) covers all the remaining terms of $T-\{t\}$.

Let T_1 and T_2 be regular sets of terms, and let t_1 and t_2 be terms that cover all of the terms of T_1 and T_2, respectively. We call the set $T_1 \cup T_2$ the regular union of T_1 and T_2 if either t_1 covers t_2 or t_2 covers t_1.

LEMMA 5. *The regular union of the regular sets T_1 and T_2 is a regular set.*

The proof follows from the transitivity of the covering relation.

LEMMA 6. *The rank of any regular term $ft_1 \cdots t_n$ is not greater than $n - m + 1$, where m is the least degree of the functional symbols occurring in $ft_1 \cdots t_n$.*

PROOF. We shall prove the lemma by induction on n. For $n = 1$ the truth of the lemma is obvious. Let $n > 1$. Since an arbitrary subterm of positive rank of the term t_i $(1 \leqslant i \leqslant n)$ has to be covered by the term $ft_1 \cdots t_n$, there must be among these subterms only terms beginning with functional symbols whose degree is

(4) The degree of an n-place functional symbol is the number n.

less than n. According to Lemma 1, every subterm of positive rank of a regular term is regular. By the induction hypothesis each term t_i of positive rank has rank not greater than $n_i - m_i + 1$, where n_i is the degree of the functional symbol with which t_i begins, and m_i is the minimal degree of the functional symbols occurring in t_i. Let the rank of t_i be maximal among the terms $t_1, \cdots, t_n$. Since the rank r of the term $ft_1 \cdots t_n$ is greater by 1 than the rank of t_i, we have

$$r \leqslant (n_i - m_i + 1) + 1 = (n_i + 1) - m_i + 1 \leqslant n - m_i + 1 \leqslant n - m + 1.$$

LEMMA 7. *Let T be a regular set of regular terms; suppose that $f_1 t_1 \cdots t_n$ and $f_2 t_1 \cdots t_m$ belong to T, that the variable v occurs in $f_1 t_1 \cdots t_n$ but not in $f_2 t_1 \cdots t_m$, and that $\{f_2 t_1 \cdots t_m / v\}$ is a substitution. Then $f_1 t_1 \cdots t_n \{f_2 t_1 \cdots t_m / v\}$ is a regular term.*

PROOF. Since v must not appear in $f_2 t_1 \cdots t_m$, it follows that n is greater than m and v is in one of the terms $t_{m+1}, \cdots, t_n$. Since T is regular, $f_1 t_1 \cdots t_n$ covers $f_2 t_1 \cdots t_m$; and since v occurs only in the terms $t_{m+1}, \cdots, t_n$, $f_1 t_1 \cdots t_n \{f_2 t_1 \cdots t_m / v\}$ covers $f_2 t_1 \cdots t_n$. Every subterm of positive rank of $f_1 t_1 \cdots t_n \{f_2 t_1 \cdots t_m / v\}$ either is an $\{f_2 t_1 \cdots t_m / v\}$-instance of some subterm of positive rank of $f_1 t_1 \cdots t_n$ or is identical to $f_2 t_1 \cdots t_m$. Since concretization preserves covering and $f_1 t_1 \cdots t_n \{f_2 t_1 \cdots t_m / v\}$ covers $f_2 t_1 \cdots t_m$, the term $f_1 t_1 \cdots t_n \{f_2 t_1 \cdots t_m / v\}$ is regular.

THEOREM 1. *Let there be given a derivation tree in $S_{\mathfrak{A}}$ with some natural ordering of the literals, and suppose that for any occurrences V_1 and V_2 of terms in any disjunctions of this tree it follows that if the rank of V_1 is greater than that of V_2 then the latent rank of V_1 is greater than the latent rank of V_2. Then in this derivation tree the set of terms of positive rank in each disjunction is regular and contains regular terms only.*

PROOF. We shall prove the theorem by induction on the number n of applications of rule P in the given derivation tree.

When $n = 0$, the rank of each term in the disjunctions of $S_{\mathfrak{A}}$ is not greater than 1, and every term of rank 1 is trivially regular; the set of terms of positive rank in each disjunction of $S_{\mathfrak{A}}$ is regular by the definition of $S_{\mathfrak{A}}$.

Let $n > 0$, let N be the root of the derivation tree, and for some C, D, L, M, ξ, η and σ let $\mathrm{P}^*(N, C, D, L, M, \xi, \eta, \sigma)$ obtain. We can distinguish two cases.

Case 1. The disjunctions C and D contain terms of positive rank. Then the principal terms of each of the premises of C and D are similar, have positive rank according to the hypothesis of the theorem, and cover all terms of positive rank in their premises because the terms of positive rank that appear in C and D are regular by the induction hypothesis.

All the terms of positive rank occurring in $C\xi\sigma$ and $D\eta\sigma$ are regular by Lemma 3. The set of terms of the disjunction N is the regular union of sets of terms of $C\xi\sigma$ and $D\eta\sigma$, from which are omitted terms occurring in $C\xi\sigma$ and $D\eta\sigma$ and not in $(C-L)\xi\sigma$

and $(D-M)\eta\sigma$. By Lemmas 5 and 4 the set of terms of N is regular and contains only regular terms.

Case 2. One of the premises, say D, does not contain terms of positive rank. Then each term of positive rank occurring in N is a concretization by means of the substitution $\{u_1/v_1, \cdots, u_k/v_k\}$ of a certain term of positive rank occurring in C, where $v_1, \cdots, v_k$ are variables appearing in D, and $u_1, \cdots, u_k$ are subterms of terms appearing in C.

Recognizing that $\{u_1/v_1, \cdots, u_k/v_k\}$ can be expressed in the form of a composition of single-element substitutions, and using Lemma 7, we obtain the result that the set of terms in N is regular and contains only regular terms.

REMARK. It is not difficult to show that for closed formulas of class K (see [2]) there exists a natural ordering of the literals such that any derivation tree with this natural ordering satisfies the conditions of Theorem 1. It follows from Lemma 6 that in these trees the ranks of the terms do not exceed a certain given number. In [5] a connection was established between the method of resolution and the inverse method which permits one to carry over results obtained by either one of the methods to the other. Using this connection, we can show that for formulas of class K the length of the chain of sets participating in a proof in which the empty set is the successful result is bounded by a number depending only on the type of the formula.

THEOREM 2. *If every term of positive rank in every literal of* $S_{\mathfrak{A}}$ *is principal, then every term of positive rank in any derivation tree in the set* $S_{\mathfrak{A}}$ *with any natural ordering is regular.*

PROOF. We shall prove that in any derivation tree in $S_{\mathfrak{A}}$ with a natural ordering each term of positive rank is principal in the literal in which it appears and has rank 1. The proof is by induction on the number k of applications of rule P in the derivation tree.

If $k=0$, then the tree contains only disjunctions from $S_{\mathfrak{A}}$, and every term of positive rank is principal in the literal in which it occurs by hypothesis. The ranks of the terms in the disjunctions of $S_{\mathfrak{A}}$ are not greater than 1 according to the definition of $S_{\mathfrak{A}}$.

Let $k>0$ and let N be the root of the derivation tree such that $P^*(N, C, D, L, M, \xi, \eta, \sigma)$ is valid for some C, D, L, M, ξ, η and σ. Every term of positive rank occurring in C or D is principal in the literal in which it appears and has rank 1 according to the induction hypothesis. If there is a component $ft_1 \cdots t_n/v$ in $\xi\sigma$ or $\eta\sigma$, then v cannot occur in any term of positive rank. Indeed, assume that v occurs in some term t having positive rank. Then the latent rank of t is greater than that of v, and t precedes v in the natural ordering. Under concretization of $(C-L)$ by $\xi\sigma$ or $(D-M)$ by $\eta\sigma$ the term $ft_1 \cdots t_n$, which is principal in its premise by the induction hypothesis, will be substituted in place of v.

Consequently v must be the principal term in its premise, since the tree being considered is a tree with natural ordering. But v cannot be a principal term since t precedes v. This contradiction proves that v cannot occur in any term of positive rank.

We have thus shown that the ranks of the terms in the tree considered cannot be greater than 1, and every term of rank 1 is trivially regular.

REMARK. One of the classes of formulas satisfying the conditions of Theorem 2 is the class of formulas that are closures of the almost-Skolem formulas defined in [2]. From this we easily deduce solvability of the class of almost-Skolem formulas $\mathfrak{A}$, and $S_{\mathfrak{A}}$ contains disjunctions of not more than two terms. Solvability of this class is established in [2].

The author is deeply grateful to G. V. Davydov and S. Ju. Maslov for their valuable advice and help during the preparation of this article for publication, and he thanks the participants in the seminar on mathematical logic at the Leningrad Branch of the V.A. Steklov Institute of Mathematics, Academy of Sciences of the USSR, for joining in the discussion of the material, as well as V. I. Šaronov for his helpful remarks.

BIBLIOGRAPHY

1. J. A. Robinson, *A machine-oriented logic based on the resolution principle*, J. Assoc. Comput. Mach. **12** (1965), 23–41; Russian transl., Kibernetika **1970**, no. 7, 194–218. MR **30** #732.

2. S. Ju. Maslov, *The inverse method for establishing deducibility for logical calculi*, Trudy Mat. Inst. Steklov **98** (1968), 26–87 = Proc. Steklov Inst. Math. **98** (1968), 25–96. MR **40** #5416; **43** #4620.

3. S. Kanger, *A simplified proof method for elementary logic*, Computer Programming and Formal Systems, North-Holland, Amsterdam, 1963, pp. 87–94; Russian transl. in Mathematical Theory of Logical Inference, "Nauka", Moscow, 1967, pp. 200–209. MR **27** #1350.

4. J. R. Slagle, *Automatic theorem proving with renamable and semantic resolution*, J. Assoc. Comput. Mach. **14** (1967), 687–697. MR **38** #3093.

5. S. Ju. Maslov, *The connection between tactics of the inverse method and the resolution method*, Zap. Naučn. Sem. Leningrad. Otdel. Mat. Inst. Steklov. (LOMI) **16** (1969), 137–146 = Sem. Math. V. A. Steklov Math. Inst. Leningrad **16** (1969), 69–73. MR **42** #1645.

Translated by R. N. GOSS

Trudy Mat. Inst. Steklov.
121 (1972)

Proc. Steklov Inst. Math.
121 (1972)

THE INVERSE METHOD AND TACTICS FOR ESTABLISHING DEDUCIBILITY FOR A CALCULUS WITH FUNCTIONAL SYMBOLS

S. Ju. MASLOV

UDC 51.01:164

ABSTRACT. A concretization of the inverse method (MR 40 #5416; 43 #4620) is presented in detail, permitting us to establish deducibility in the classical predicate calculus (for assumption formulas of arbitrary structure). Completeness theorems are proved for a series of tactics (in various combinations), which permits rigid regulation of verifications in the process of establishing deducibility; special attention is given to tactics based on unification of the order of terms in favorable sequences. A connection is established between complexities of the basis of favorability and Gentzen inference (for complexity of the latter a number of applications of ∃-rules are admitted).

Bibliography: 24 items.

CONTENTS. Introduction. §1. Formulation of the method for standardized sequents. §2. Completeness of the method and properties of favorable sequences. §3. Π-tactics. §4. Tactics combined with the tactic of supports. §5. The general case. Decompositions with repetitions. §6. Generalization of rule Б. Completeness of the method. §7. Extension of tactics. §8. Decompositions without repetitions. Weak Π-tactics. Bibliography.

Introduction

The aim of the present paper is a detailed presentation of the inverse method for establishing deducibility (cf., for example, [1]) for the classical predicate calculus with functional symbols (without equality), as well as a study of tactics enabling us to restrict the verification process in work based upon the inverse method. The proposed concretization of the method (the most general form of which is presented only in §8.1) is intended to serve as a basis for practical machine algorithms for establishing deducibility. In this connection, it is clear that, within the framework of the proposed concretization, one can formulate various algorithms, differing in purpose, complexity, and practical effectiveness.

Since the requirement of reducing the formula to be tested to prenex or some other standard form generally decreases the practical effectiveness of the method, the

---

AMS (MOS) subject classifications (1970). Primary 02B10, 02G99, 68A40.

proposed concretization will apply to formulas of arbitrary structure. (We shall use only a preliminary standardization of formulas which does not increase the difficulty of obtaining a derivation.) In this connection, the basic definitions turn out to be quite tedious. Therefore the first four sections are devoted to a simpler concretization, the basic features (and the basic ideas of the proofs of the results) of which are quite easy to carry over to the general case of sequents of arbitrary structure.

The tactics of the inverse method have much in common with the tactics of the method of resolution (cf., for example, [2]–[5]). The connection between these methods [6]–[8] is very useful for their joint application and for a comparative study of the tactical possibilities of establishing deducibility; in this paper we shall discuss the method of resolution only to clarify the correspondence between tactics.

Some results of this paper have been formulated in [6], [7], [9]. [9] also contains a sketch of a proof of a simplified variant of Theorem 2. For clarification of the "intuitive meaning" of certain concepts which we shall introduce, familiarity with Chapter 1 of [1] would be useful; it is necessary for the reader to know the methods of sequents (for example, [10], [11]).

The numbering of theorems, lemmas, remarks, and displayed expressions is consecutive throughout the entire text of the paper.

The author considers it an agreeable duty to express his gratitude to V. P. Orevkov and Ju. V. Matijasevič, with whom he had useful discussions about the content and form of this paper.

§1. Formulation of the method for standardized sequents

1. We shall establish the deducibility of pure sequents of the form

$$\to \exists x_{1,1} \ldots x_{1,m_1}\left(\mathop{\&}_{j=1}^{\delta_1} D_{1,j}\right), \ldots, \exists x_{n,1} \ldots x_{n,m_n}\left(\mathop{\&}_{j=1}^{\delta_n} D_{n,j}\right), \tag{1}$$

where $D_{1,1}, \cdots, D_{n,\delta_n}$ are simple disjunctions(1) in a language with functional symbols. By deducibility we shall understand deducibility in the following calculus, which, on sequents of form (1), is equivalent to a familiar variant of the classical predicate calculus with functional symbols. Below, D and D' are simple disjunctions; Γ, Γ' and Γ'' are lists of formulas (possibly empty). The axiom schemas of the calculus are the following:

1) $\to \Gamma, D, \Gamma'$, where D is closed;

2) $\to \Gamma, D, \Gamma', D', \Gamma''$, where $D \vee D'$ is closed.

There are two rules of inference:

---

(1) A simple disjunction is a disjunction of elementary formulas which are atomic formulas or negations of atomic formulas; a disjunction is said to be *closed* if it contains a formula together with its negation.

$$\frac{\to \Gamma,\ B^1,\ \Gamma';\ \ldots;\ \to \Gamma,\ B^s,\ \Gamma'}{\to \Gamma,\ B^1 \,\&\, \ldots \,\&\, B^s,\ \Gamma'}\ (\to \&),$$

where $B^1, \cdots, B^s$ do not have & as principal symbol;

$$\frac{\to \Gamma,\ [M]^{x_1, \ldots, x_m}_{t_1, \ldots, t_m},\ \exists x_1 \ldots x_m M,\ \Gamma'}{\to \Gamma,\ \exists x_1 \ldots x_m M,\ \Gamma'}\ (\to \exists),$$

where M is quantifier-free, and $[M]^{x_1, \ldots, x_m}_{t_1, \ldots, t_m}$ denotes the result of substituting for the variables $x_1, \cdots, x_m$ in M the terms $t_1, \cdots, t_m$, each of which does not contain bound variables of the succedent.

We denote this calculus by $\mathbf{G}_0$. Let us agree that the sequent (1) is to be considered closed; that is, any free variables in it are to be considered not as variables but as functional constants. Without restricting generality, we shall assume that, for any i, j, i' and j' $(1 \leqslant i \leqslant n;\ 1 \leqslant i' \leqslant n;\ 1 \leqslant j \leqslant \delta_i;\ 1 \leqslant j' \leqslant \delta_{i'};\ (i - i')^2 + (j - j')^2 \neq 0)$ it is not true that $D_{i,j}$ occurs in $D_{i',j'}$ (in particular, all the $D_{i,j}$ are distinct). We note that it is easy to satisfy this restriction by applying certain equivalences which simplify the problem of establishing deducibility (for example, the equivalences $D \equiv D \,\&\, (D \vee D')$ and $\exists x(D \,\&\, M) \equiv D \,\&\, \exists x M$ for D not containing x, etc.). As a result of applying a chain of such equivalences, for any sequent (1) we are able to construct certain simpler sequents which satisfy the restriction just mentioned and whose simultaneous deducibility is equivalent to the deducibility of sequent (1).

2. Let us denote by F_i the ith formula of sequent (1) $(i = 1, \cdots, n)$, and by F the formula $F_1 \vee \cdots \vee F_n$. By the *decomposition* of the formula F and sequent (1) we mean the list

$$D_{1,1}, \ldots, D_{1,\delta_1}, \ldots, D_{n,1} \ldots, D_{n,\delta_n}; \tag{2}$$

the total number of terms of this list will be denoted by δ. (Free) variables of the formula $D_{i,j}$ will be called *basic* variables of this formula (remember that all of them will be contained in the list $x_{i,1}, \cdots, x_{i,m_i}$). Let C be any list of natural numbers

$$\alpha_1, \alpha_2, \ldots, \alpha_h \quad (h \geqslant 0), \tag{3}$$

not exceeding δ, and let

$$W_1, W_2, \ldots, W_h \tag{4}$$

be the list composed of the α_1th, $\cdots$, α_hth formulas of the decomposition (2). By an (F, C)-*atom* (briefly, *atom*) we mean any expression of the form u^k, where u is a basic variable of the formula W_k $(1 \leqslant k \leqslant h)$. Terms which contain functional symbols only from F (and, possibly, arbitrary variables) will be called *F-terms.* By an (F, C)-*bundle* (or *bundle*) we mean a list of expressions $A \approx T$ (where A is an

(F, C)-atom and T is an F-term) such that every (F, C)-atom is mentioned in the list at most once, each of the expressions in the list is called a *relation* of the bundle. By variables of the bundle we mean only variables occurring in F-terms of the bundle. By a *system of values* of (F, C)-atoms we mean an object of the form

$$\begin{pmatrix} A_1, \dots, A_\varepsilon \\ T_1 \dots, T_\varepsilon \end{pmatrix}, \tag{5}$$

where $A_1, \cdots, A_\epsilon$ is a complete list without repetitions of all (F, C)-atoms, and $T_1, \cdots, T_\epsilon$ are arbitrary terms, called the *values* of the corresponding atoms (in the system (5)). We shall say that (5) *satisfies* an (F, C)-bundle S if for every variable of S one can find a term such that, upon substitution of these terms in S, every relation in S is transformed into an expression of the form $A_i \approx T_i$ $(1 \leqslant i \leqslant \epsilon)$.

EXAMPLE. Let us denote by F^0 the formula

$$\exists xyz \frac{Q(fx, fy) \vee \neg Q(x, fx) \vee \neg Q(fy, y)}{P(fx, fy, fz) \vee \neg P(fy, fx, z)},$$

where a horizontal line is used instead of the symbol &. The decomposition of the sequent $\to F^0$ is the list

$$Q(fx, fy) \vee \neg Q(x, fx) \vee \neg Q(fy, y), \ P(fx, fy, fz) \vee \neg P(fy, fx, z).$$

Let us denote the list 1, 2 by C^0. The list $x^1 \approx v$, $y^1 \approx fv$, $y^2 \approx fv$ is an example of an (F^0, C^0)-bundle; the only variable of this bundle is the variable v. This bundle is satisfied, for example, by the following system of values of (F^0, C^0)-atoms:

$$\begin{pmatrix} x^1, & y^1, & z^1, & x^2, & y^2, & z^2 \\ fa, & ffa, & fb, & b, & ffa, & fa \end{pmatrix}.$$

It is convenient to use $y^1 \approx fx^1 \approx y^2$ as an abbreviation for the bundle just described. We shall use abbreviations of this kind in what follows (but only in the simplest cases not admitting of any ambiguity, and only in dealing with examples). Thus the expression $x^1 \approx fx^2$, $y^2 \approx z^2 \approx ffy^1$ is used as an abbreviation for the (F^0, C^0)-bundle $x^1 \approx fv$, $x^2 \approx v$, $y^1 \approx w$, $y^2 \approx ffw$, $z^2 \approx ffw$.

REMARK 1. The concept of bundle that we have introduced permits a significant simplification in our exposition (primarily because any system of relations which can be written down in the form of a bundle is admissible in the sense of [1], §2.3); however, this concept is very uneconomical from a practical point of view. The choice of language for presenting relations is, fortunately, reflected only in the following definitions, but not at all in the results or proofs, whose content is independent of this language (almost all the results below). For comparison with the concept of bundle in [12] (or [1], §6.1), we remark that the variables and functional symbols occurring in our bundle play the roles, respectively, of the expressions $\lambda, \lambda', \lambda'', \cdots$ and the symbols $g^i_{1,\dots,K_i}$ (these symbols, which we call *total*, suffice for the expression of arbitrary relations; cf. [1], footnote 6). For example, a relation of the form $A \approx g^i_{1,3}(\lambda^{(s)}, A')$, where A and A' are atoms and g_i is a ternary functional symbol, is written now in the form $A' \approx r, A \approx g_i(s, t, r)$, where t is a variable which occurs nowhere else in the bundle, and s and r are variables which, in all F-terms of the bundle, are used instead of the expression $\lambda^{(s)}$ and the atom A', respectively. The introduction of nontotal symbols might have certain practical advantages in certain respects and from a theoretical point of view (cf. footnote 9 below). The way of presenting relations which is most suitable for machine algorithms is the method described in [13].

3. Any pair of the form

$$[\mathrm{C}; S] \text{ (i.e. } [\alpha_1, \alpha_2, \dots, \alpha_h; S]), \tag{6}$$

where C is a list of the form (3) and S is an (F, C)-bundle, will be called an F-*sequence* (or *sequence*). The number h will be called the *length* of the sequence. A sequence of length zero will be called *empty*.

Let us consider some branch **B** of the proof search tree(2) for the sequent (1). Assume that h figures occur in **B**, the kth of which looks as follows:

$$\cfrac{\cfrac{\dots; \to \Gamma, \overline{W}_k, \exists x_1 \dots x_m (K \,\&\, W_k \,\&\, K'), \Delta; \dots}{\to \Gamma, (\overline{K} \,\&\, \overline{W}_k \,\&\, \overline{K}'), \exists x_1 \dots x_m (K \,\&\, W_k \,\&\, K'), \Delta}\,(\to \&)}{\to \Gamma, \exists x_1 \dots x_m (K \,\&\, W_k \,\&\, K'), \Delta}\,(\to \exists),$$

where W_k has the same meaning as in (4). A line over a symbol denotes the result of the appropriate substitution in the formula denoted by this symbol. We construct a system of values of (F, C)-atoms, including as a value of the atom u^k the term which is substituted for u in the middle sequent of the displayed figure (u is a basic variable of the formula W_k). If this system of values satisfies S, then we shall say that the sequence (6) occurs in **B**, that the upper sequents of the figure above are the sequents containing (6), and that the list of the indicated occurrences of formulas $\overline{W}_1, \cdots, \overline{W}_h$ are *realizations* of the sequence (6) in **B**. (A given sequence can have several realizations in a branch.) The following is obvious.

LEMMA 1. *For any F-sequence one can construct a branch of the proof search tree in which it occurs.*

Continuation of the example. In the branch

---

(2) A proof search tree differs from a proof tree in that its uppermost sequents need not be axioms.

$$\dfrac{\dfrac{\dfrac{\dfrac{\ldots;\to P(fa,fb,fc)\vee\neg P(fb,fa,c),\ P(fb,fa,ffc)\vee\neg P(fa,fb,fc),\ F^0}{\to P(fa,fb,fc)\vee\neg P(fb,fa,c),\ \dfrac{Q(fb,fa)\vee\neg Q(b,fb)\vee\neg Q(fa,a)}{P(fb,fa,ffc)\vee\neg P(fa,fb,fc)},\ F^0}\,(\to\&)}{\ldots;\to P(fa,fb,fc)\vee\neg P(fb,fa,c),\ F^0}\,(\to\exists)}{\to\dfrac{Q(fa,fb)\vee\neg Q(a,fa)\vee\neg Q(fb,b)}{P(fa,fb,fc)\vee\neg P(fb,fa,c)},\ F^0}\,(\to\&)}{\to F^0}\,(\to\exists)$$

the F^0-sequence $[2, 2; x^1 \approx y^2, x^2 \approx y^1, z^2 \approx fz^1]$ occurs.

We shall say that a sequence $\mathbf{H}$ of the form (6) is obtained from a sequence $\mathbf{H}'$ of the form $[\alpha'_1, \cdots, \alpha'_{h'}; S']$ by *excision with respect to the characteristic* $j_1, \cdots, j_h$ (here $h \leqslant h'$ and $j_1, \cdots, j_h$ are numbers such that $1 \leqslant j_1 < j_2 < \cdots < j_h \leqslant h'$) if:

1) $\alpha'_{j_1} = \alpha_1, \alpha'_{j_2} = \alpha_2, \cdots, \alpha'_{j_h} = \alpha_h$; or

2) S is obtained from S' by eliminating all relations of the form $u^k \approx T$, where $k \notin \{j_1, \cdots, j_h\}$ and, after this is done, replacing every relation of the form $u^{j_k} \approx T$ $(k = 1, \cdots, h)$ by the relation $u^k \approx T$.

If, in addition, the characteristic has the form $1, \cdots, h$, then $\mathbf{H}'$ will be called an *extension* of the sequence $\mathbf{H}$, and if, in addition, $h' = h + 1$, we shall say that $\mathbf{H}'$ is an *elementary extension* of $\mathbf{H}$.

The sequence (6) will be called a *subsequence* of a sequence $[C; S']$ with respect to the characteristic $1, 2, \cdots, h$ if every variable of S can be replaced by a term in such a way that every relation of S is transformed into some relation of S'. If the indicated replacement is only a renaming of the variables, the sequences will be said to be *equal.* We shall say that a sequence $\mathbf{H}$ of the form (6) is a *subsequence* of a sequence $\mathbf{H}'$ of the form $[\alpha'_1, \cdots, \alpha'_{h'}; S']$ *with respect to the characteristic* $j_1, \cdots, j_h$ if $\mathbf{H}$ is a subsequence with respect to the characteristic $1, \cdots, h$ of some sequence obtained from $\mathbf{H}'$ by excision with respect to the characteristic $j_1, \cdots, j_h$. If the characteristic is such that $j_h = h'$, then $\mathbf{H}$ will be said to be a subsequence *cofinal* with $\mathbf{H}'$. If one can find a characteristic with respect to which $\mathbf{H}$ is a subsequence of $\mathbf{H}'$, then $\mathbf{H}$ will be called a *subsequence* of $\mathbf{H}'$ and $\mathbf{H}'$ will be called a *supersequence* of $\mathbf{H}$; if, moreover, $\mathbf{H}$ is not equal to $\mathbf{H}'$, then we shall say that $\mathbf{H}$ is a *proper subsequence* of $\mathbf{H}'$ and that $\mathbf{H}'$ is a *proper supersequence* of $\mathbf{H}$.

LEMMA 2. *Let* $\mathbf{H}$ *be a subsequence of* $\mathbf{H}'$ *with respect to the characteristic* $j_1, \cdots, j_h$. *Then, in every branch in which* $\mathbf{H}'$ *occurs,* $\mathbf{H}$ *occurs also. Moreover, if the sequents containing* $\mathbf{H}'$ *are* $\Sigma_1, \cdots, \Sigma_{h'}$, *then the sequents containing* $\mathbf{H}$ *are* $\Sigma_{j_1}, \cdots, \Sigma_{j_h}$.

The proof of the lemma is an immediate consequence of the following obvious assertions: 1) If $\mathbf{H}$ is obtained from $\mathbf{H}'$ by excision, then it occurs in all branches containing $\mathbf{H}'$ (and thus the required relation in the lemma between the sequents containing these sequences is fulfilled). 2) If [C, S] is a subsequence of [C, S'], then any system of values of (F, C)-atoms satisfying S' also satisfies S.

Bundles S and S' are said to be *inconsistent* if, for some atom A, $A \approx T$ occurs in S, and $A \approx T'$ occurs in S', but $T \neq T'$. A collection of elementary extensions

$$[C, \beta; S_1], \quad [C, \beta+1; S_2], \ldots, [C, \beta+\nu-1; S_\nu] \qquad (7)$$

of the sequence (6) is said to be *complete* if $S_1, \cdots, S_\nu$ are pairwise consistent and, for some i $(1 \leqslant i \leqslant n)$, β is the index of the formula $D_{i,1}$ in the decomposition (2) and $\nu = \delta_i$.

Continuation of the example. The F^0-sequences $[1, 2, 1; y^1 \approx fx^1 \approx y^2 \approx x^3, x^2 \approx y^3]$ and $[1, 2, 2; y^1 \approx fx^1 \approx y^2 \approx x^3, x^2 \approx y^3, z^3 \approx fz^2]$ form a complete collection of extensions of the sequence $[1, 2; y^1 \approx fx^1 \approx y^2]$.

LEMMA 3. *Let* (7) *be a complete collection of elementary extensions of the sequence* (6). *Then every branch in which* (6) *occurs can be built upwards (in accordance with the rules of* $\mathbf{G}_0$) *in such a way that one obtains* ν *branches, in each of which one of the sequences of* (7) *occurs.*

Let (6) occur in **B** and let Ω be a system of values of some sequence occurring in the branch. Since Ω satisfies S, for every variable of S one can find terms such that, if they are substituted in the relations of S, we obtain values of the atoms of Ω. If there are in $S_1, \cdots, S_\nu$ no variables other than the ones mentioned, then all atoms of these bundles are obtained after suitable substitutions of certain values. If some variables of $S_1, \cdots, S_\nu$ are not mentioned in S, one can substitute for them arbitrary terms not containing $x_{1,1}, \cdots, x_{n,m_n}$. We thus arbitrarily assign values to atoms with upper index $h+1$ not mentioned in $S_1, \cdots, S_\nu$. As a result, we obtain ν systems; in the lth of these systems there are values of (F, (C, $\beta + l$))-atoms $(l = 0, 1, \cdots, \nu - 1)$. On ($F$, C)-atoms all systems coincide with Ω. Because of the pairwise consistency of the bundles there are also no inconsistencies among these systems. Using the resulting values of atoms with upper index $h+1$, we build onto **B** by "counterapplication" of ($\rightarrow \exists$) (with the formula F_i principal), with substitution of these values(3) for the variables, and then by means of counterapplication of ($\rightarrow$ &) to the resulting conjunction. It is easily verified that $[C, \beta + l - 1; S_l]$ occurs in the lth resulting branch $(l = 1, \cdots, \nu)$.

4. By the formula *corresponding to the sequence* (6) we mean the formula $\overline{W}_1 \vee \cdots \vee \overline{W}_h$, where $\overline{W}_k$ is obtained from W_k (in the list (4)) by substitution of the term T for the variable u if the relation $u^k \approx T$ is contained in S, and some

(3) The restriction on terms to be substituted turns out to be fulfilled. If F_i does not contain fictitious variables, then the values of all necessary terms are constructed, and the value of a fictitious variable is not important.

new variables (different from each other and those occurring in S) in contrary cases; all such formulas are mutually congruent. We shall call the sequence *closed* if the corresponding disjunction is closed.

Now we shall define by induction the concept of *favorable sequence.*

RULE **A**. Any closed sequence is considered favorable.

RULE **Б**. Assume that $\mathbf{H}_1, \cdots, \mathbf{H}_\nu$ are favorable sequences and that the sequence (6) possesses the following property: one can find a complete collection $\mathbf{H}'_1, \cdots, \mathbf{H}'_\nu$ of elementary extensions of the sequence (6) such that, for all i $(1 \leqslant i \leqslant \nu)$, $\mathbf{H}_i$ is a cofinal subsequence of $\mathbf{H}'_i$. Then (6) is considered a favorable sequence.(4)

Applicability of Rules **A** and **Б** (to given arguments) is decidable. Moreover, one can assign to these rules an operator form which is convenient for application. In the sequel we shall indicate a series of possibilities of sharp restriction on the verifications necessary in applications of **A** and **Б**. In the meantime, what is important for us is that, organizing according to these rules the process of constructing favorable sequences, aimed toward obtaining the empty sequence, we arrive at a complete and correct method for establishing deducibility in the calculus $\mathbf{G}_0$. In fact, in the following section we shall prove

THEOREM 1. *A sequent of the form* (1) *is derivable in the calculus* $\mathbf{G}_0$ *if and only if the empty F-sequence is favorable.*

Conclusion of the example. The F^0-sequences

$$[1,\ 1;\ y^1 \approx fx^1 \approx x^2],$$
$$[1,\ 1;\ x^1 \approx fy^1 \approx y^2],$$
$$[2,\ 2;\ x^1 \approx y^2;\ y^1 \approx x^2,\ z^2 \approx fz^1]$$

are closed (see the branch constructed above, in which $\mathbf{H}_3$ occurs). Let us consider a complete collection of extensions of the sequence

$$[1,\ 2;\ y^1 \approx fx^1 \approx y^2]$$

(see the continuation of the example on p. 17). Since the first sequence of this collection is a supersequence of $\mathbf{H}_1$, and the second is a supersequence of $\mathbf{H}_3$ (the corresponding characteristics are 1, 3 and 2, 3), $\mathbf{H}_4$ is then a result of applying **Б** to $\mathbf{H}_1$ and $\mathbf{H}_3$. Similarly (with $\mathbf{H}_2$ playing the role of $\mathbf{H}_1$) we establish that the sequence

$$[1,\ 2;\ x^1 \approx fy^1 \approx x^2]$$

is favorable.

The sequences $\mathbf{H}_1$ and $\mathbf{H}_4$ (respectively, $\mathbf{H}_2$ and $\mathbf{H}_5$) form a complete collection of extensions of the sequence

(4) The last components of the sequences H_i (and H'_i) will be called the *outer components* of the application of **Б**.

$$[1;\ y^1 \approx fx^1]$$

and respectively, of the sequence

$$[1;\ x^1 \approx fy^1].$$

Therefore $\mathbf{H}_6$ and $\mathbf{H}_7$ are favorable. Applying **Б** to $\mathbf{H}_3$ and $\mathbf{H}_6$, we have

$$[2;\ x^1 \approx fy^1].$$

Finally, applying **Б** to $\mathbf{H}_7$ and $\mathbf{H}_8$, we obtain the empty F^0-sequence. By Theorem 1 the sequent $\to F^0$ is derivable in $\mathbf{G}_0$.

§ 2. Completeness of the method and properties of favorable sequences

1. Let **B** be a branch of a proof search tree of sequent (1) in which: 1) applications of the rules $(\to \exists)$ and $(\to \&)$ alternate (a side formual of an application of $(\to \exists)$ is the principal formula of the next higher application of $(\to \&)$); 2) in counterapplications of $(\to \exists)$, all the terms which are substituted are F-terms. We associate with the branch **B** a certain sequence **H(B)**, which we call *equivalent* to this branch. The length of the sequence **H(B)** is equal to the number of applications of $(\to \&)$ in **B**; the kth component of the list of the sequence **H(B)** is the index of the formula of the decomposition from which by substitution one obtains the side formula of the kth from below application of $(\to \&)$; for every variable u of this formula, there is included in the bundle of the sequence **H(B)** the relation $u^k \approx T$, where T is substituted for u in the application of $(\to \exists)$ lying immediately below the given application of the rule $(\to \&)$. **H(B)** occurs in **B** but no proper subsequence of it occurs in **B**.

Assume that the sequent (1) is derivable and that $\mathfrak{D}$ is its proof tree, composed of branches as in the preceding paragraph (any tree is easily transformed into such a tree). It is evident that each of the branches of this tree is equivalent to a closed sequence. Let us consider an arbitrary application of the rule $(\to \&)$ in $\mathfrak{D}$ and let **B** be a branch of the tree leading to the lower sequent of this application (and breaking off at this sequent). Let $\mathbf{B}_1, \cdots, \mathbf{B}_\nu$ be the branches obtained by building **B** one level higher (ν is the number of premises of the given application of rule $(\to \&)$). It is evident that by one application of rule **Б** to sequences equivalent to $\mathbf{B}_1, \cdots, \mathbf{B}_\nu$ one can obtain a sequence equivalent to **B**. Using this assertion in the induction step, we obtain (by induction on the number of applications of $(\to \&)$ in $\mathfrak{D}$) the fact that the empty F-sequence is favorable (compare [1], § 8.4). Hence the method described in § 1 is complete.

2. The proofs of the correctness of the method and other results of this section are carried out in a way completely analogous to that in § 9 of [1]. Therefore we shall present only the statements here.

Assume that $\mathbf{H}$ and $\mathbf{H}_1, \cdots, \mathbf{H}_f$ are arbitrary F-sequences. Any list of F-sequences terminating with the sequence $\mathbf{H}$ is said to be a *foundation for the favorability of* $\mathbf{H}$ with respect to $\mathbf{H}_1, \cdots, \mathbf{H}_f$ if every sequence $\mathbf{H}'$ of this list is either closed or is obtained by Rule **Б** from sequences lying in the list to the left of $\mathbf{H}'$ or coincides with one of the sequences $\mathbf{H}_1, \cdots, \mathbf{H}_f$. When a corresponding foundation exists, we shall say that $\mathbf{H}$ is favorable with respect to $\mathbf{H}_1, \cdots, \mathbf{H}_f$. For $f = 0$, the words "with respect to" are omitted. (In the sequel we shall resort to a tree-like way of writing foundations.)

LEMMA 4. *Let an F-sequence $\mathbf{H}$ be favorable with respect to $\mathbf{H}_1, \cdots, \mathbf{H}_f$, and let $\mathbf{B}$ be a branch in which $\mathbf{H}$ occurs. One can build onto $\mathbf{B}$ in such a way that in each of the resulting branches either a closed sequence or one of the sequences $\mathbf{H}_1, \cdots, \mathbf{H}_f$ occurs.*

COROLLARY. *Let $\mathbf{H}$ be favorable and let $\mathbf{B}$ be a branch in which it occurs. Then the uppermost sequent of $\mathbf{B}$ is derivable. In particular, if $\mathbf{H}$ is empty, then the sequent* (1) *is derivable.*

Thus *the method formulated in* §1 *is correct.* This completes the proof of Theorem 1.

We shall say that the sequence (6) is *factorable* if one can find h' $(1 < h' \leqslant h)$ such that, excising from (6) sequences $[\mathrm{C}_1; S_1]$ and $[\mathrm{C}_2; S_2]$ with respect to the characteristics $1, \cdots, h' - 1$ and $h', h' + 1, \cdots, h$, respectively, we obtain S_1 and S_2, not having variables in common.(5) Moreover, we say that (6) *factors* into $[\mathrm{C}_1; S_1]$ and $[\mathrm{C}_2; S_2]$. We say that a sequence $\mathbf{H}$ *factors* into the sequences $\mathbf{H}^1, \cdots, \mathbf{H}^t$ if $\mathbf{H}$ factors into $\mathbf{H}^1$ and $\mathbf{H}'$, $\mathbf{H}'$ factors into $\mathbf{H}^2$ and $\mathbf{H}''$, $\cdots$, and $\mathbf{H}^{(t-2)}$ factors into $\mathbf{H}^{t-1}$ and $\mathbf{H}^t$. In this connection, each of the sequences $\mathbf{H}^1, \cdots, \mathbf{H}^t$ will be called a *factor* of $\mathbf{H}$.

LEMMA 5. *Assume that $\mathbf{H}$ factors into $\mathbf{H}^1, \cdots, \mathbf{H}^t$ and that $\mathbf{H}_1, \cdots, \mathbf{H}_f$ are arbitrary sequences. If an empty sequence is favorable with respect to each of the groups of sequences $\mathbf{H}^i, \mathbf{H}_1, \cdots, \mathbf{H}_f$ $(i = 1, \cdots, t)$, then it is also favorable with respect to the sequences $\mathbf{H}, \mathbf{H}_1, \cdots, \mathbf{H}_f$.*

For the proof of this lemma in [1], §9, one uses the operation of writing an F-sequence $[\mathrm{C}_1; S_1]$ to the left of the F-sequence $[\mathrm{C}_2; S_2]$. For the language considered here, this operation is defined in the following way:

1) Every atom u^k of S_2 is replaced by the atom u^{k+h_1}, where h_1 is the length of C_1.

---

(5) Remember that only variables mentioned in F-terms of a bundle are considered variables of the bundle, while free variables of the formulas are considered functional constants.

2) Variables common to S_1 and S_2 are replaced in S_1 by new variables not occurring in S_1 or S_2.

3) The bundles, transformed as described above, are joined into an $(F,(C_1, C_2))$-bundle S, and the result of this operation is defined to be the sequence $[C_1, C_2; S]$.

LEMMA 6. *For any F-sequences* $\mathbf{H}_1, \cdots, \mathbf{H}_f$ *and a subsequence* $\mathbf{H}'_1$ *of the sequence* $\mathbf{H}_1$, *an F-sequence favorable with respect to* $\mathbf{H}_1, \cdots, \mathbf{H}_f$ *will have a subsequence favorable with respect to* $\mathbf{H}'_1, \mathbf{H}_2, \cdots, \mathbf{H}_f$.

From this lemma there arises the possibility of a sharp restriction on the verifications involved in the construction of favorable sequences, and, in particular, the appropriateness of a modification of Rules **A** and **Б**. Rules $\mathbf{A}^+$ and $\mathbf{Б}^+$ are obtained by adding to the formulation of the corresponding rules the requirement that no proper subsequence of the result of applying the rule possesses the property indicated in the rule. Results of applying Rule $\mathbf{A}^+$ will be called +-closed sequences (all such sequences have no more than two components). Rules $\mathbf{A}^+$ and $\mathbf{Б}^+$ are especially easy to transform into operator form.

3. We shall say that a sequence $[C'; S']$ is obtained from (6) as a result of applying a transposition operation if one can find numbers k and l $(1 \leqslant k \leqslant h$, $1 \leqslant l \leqslant h, l \neq k)$ such that C' is obtained from C by exchanging the kth and lth terms of the list, and S' is obtained from S as a result of simultaneous replacement of every (F, C)-atom of the form u^k (respectively, of the form u^l) by the (F, C)-atom u^l (respectively, by u^k).

We say that a sequence $\mathbf{H}'$ is a *permutation* of a sequence $\mathbf{H}$ if it can be obtained from $\mathbf{H}$ as a result of a sequence of transposition operations. We shall say that a sequence $[\alpha_1, \cdots, \alpha_{h-1}; S']$ is obtained from (6) as a result of an application of a *contraction operation* if the following conditions hold:

1) $\alpha_h = \alpha_{h-1}$.[6]

2) For every variable u of S one can find an F-term T_u such that, after substitution of these terms in S, replacement of every atom of the form w^h by the atom w^{h-1}, and elimination of repeated relations, we obtain S'; that is, as a result of the indicated changes there cannot arise two relations $w^{h-1} \approx T$ and $w^{h-1} \approx T'$ with different T and T'.

3) No proper subsequence of the resulting sequence possesses properties 1) and 2).

The result of applying a contraction operation is uniquely determined up to equality of sequences. We shall say that a sequence $\mathbf{H}_c$ is a *specification* of the sequence (6) with respect to the characteristic

$$\tau_1, \tau_2, \ldots, \tau_h \tag{8}$$

---

[6] The meaning of α_h is as in list (3).

($\tau_1, \cdots, \tau_h$ are arbitrary natural numbers; the total number of distinct numbers in this list turns out to be the length of the sequence $\mathbf{H}_c$) if it can be obtained from (6) by the following sequence of operations (we shall say that the kth component of sequence (6) is characterized by the number τ_k):

1) The left-most term of the sequence (6) characterized by the number 1 is moved to the right-hand end of the sequence, then the following term characterized by this number, etc. (the number of terms moved will be denoted by s_1, $s_1 \geqslant 0$).

2) If $s_1 > 1$, we carry out the contraction operation $s - 1$ times.

3) We move to the right-hand side the terms characterized in (6) by the number 2 (the number of such terms is s_2, $s_2 \geqslant 0$).

4) We carry out $s_2 - 1$ contractions, etc.

In the formulation of this definition, we precisely specified the sequence of permutations and contractions leading to $\mathbf{H}_c$. It is evident that this sequence has no essential significance; more precisely, it is not difficult to give an equivalent (quite complicated) definition of specification with respect to a characteristic in which this sequence of operations is not mentioned. We say that a sequence $\mathbf{H}_c$ is a *specification* of a sequence $\mathbf{H}$ if it can be obtained from $\mathbf{H}$ as a result of some sequence of permutations and contractions. The following is obvious.

LEMMA 7. *If* $\mathbf{H}'$ *is a subsequence of sequence* (6) *with respect to the characteristic* $j_1, j_2, \cdots, j_h$, *and* $\mathbf{H}_c$ *is a specification of sequence* (6) *with respect to the characteristic* (8), *then there is a specification of sequence* $\mathbf{H}'$ *with respect to the characteristic* $\tau_{j_1}, \cdots, \tau_{j_h}$ *which is a subsequence of sequence* $\mathbf{H}_c$.

Now one easily obtains the following lemma.

LEMMA 8. *Any specification of a favorable sequence is favorable.*(7)

The proof proceeds by induction with respect to the length of the foundation for favorability. The base of the induction, the fact that a specification of a closed sequence is closed, is proved by induction on the number of contraction operations used in obtaining the specification (permutations obviously have no effect on the property of being closed). Assume that $\mathbf{H}_c$ is obtained from a closed sequence (6) by a contraction. To the sequence (6) there corresponds some formula $X_1 \vee \cdots \vee X_h$, and to the sequence $\mathbf{H}_c$ a formula $X_1^* \vee \cdots \vee X_{h-1}^*$. By the construction of $\mathbf{H}_c$, the formula $X_1^* \vee \cdots \vee X_{h-1}^* \vee X_{h-1}^*$ can be obtained from $X_1 \vee \cdots \vee X_{h-1} \vee X_h$ as a result of some replacement of variables by terms. Therefore $X_1^* \vee \cdots \vee X_{h-1}^* \vee X_{h-1}^*$ is closed, and from the fact that a formula of the form $X \vee Y \vee Y$ is closed it obviously follows that $X \vee Y$ is closed.

(7) The favorableness of any permutation of a favorable sequence was proved by V. P. Orevkov.

The inductive step consists of the proof that, for every application of Rule **Б** and every specification $\mathbf{H}_c$ of the sequence which results, one can construct a specification of sequence arguments from which, in one application of Rule **Б**, one can derive $\mathbf{H}_c$. Assume that $\mathbf{H}_c$ is a specification of sequence (6) with respect to characteristic (8), and assume that $\mathbf{H}_1, \cdots, \mathbf{H}_\nu$ and $\mathbf{H}'_1, \cdots, \mathbf{H}'_\nu$ are as in the formulation of Rule **Б** (§1.4). We construct specifications of sequences $\mathbf{H}'_1, \cdots, \mathbf{H}'_\nu$, with respect to characteristic $\tau_1, \cdots, \tau_h, \max_{1 \leqslant i \leqslant h}\{\tau_i\} + 1$ (and we denote them by $\mathbf{H}'_{1c}, \cdots, \mathbf{H}'_{\nu c}$). Since the bundles $S_1, \cdots, S_\nu$ have almost all variables in common and the variables in which they differ occur only in relations for atoms with upper index $h + 1$, it follows that, in the transition to specifications, corresponding bundles also will be pairwise consistent. Thus the sequences $\mathbf{H}'_{1c}, \cdots, \mathbf{H}'_{\nu c}$ constitute a complete collection of extensions of the sequence $\mathbf{H}_c$. Therefore **Б** is applicable to any of their cofinal subsequences, with result $\mathbf{H}_c$. Lemma 7 is used (for every i, the role of the sequence $\mathbf{H}'$ is played by $\mathbf{H}_i$, sequence (6) is the sequence $\mathbf{H}'_i$, and the specification $\mathbf{H}_c$ is the sequence $\mathbf{H}'_{ic}$; $i = 1, \cdots, \nu$). We obtain specifications of the sequences $\mathbf{H}_1, \cdots, \mathbf{H}_\nu$ which are the required arguments of a new application of Rule **Б**.

From the lemma there arises the possibility of obtaining, as a result of the construction of favorable sequences, only the most "decomposed" variants of sequences which are results of applying Rule **Б** and, from all conceivable permutations of the sequences, choosing only one, after which we can use, as we desire, any specifications of these "decomposed" sequences. Sometimes a specification of a sequence can turn out to be one of its subsequences; such a specification will be called a trivial specification. From Lemmas 6 and 8 it follows that, in the process of constructing favorable sequences from a sequence admitting a trivial specification, one can immediately pass to this specification. In this connection, everywhere below where an application of **Б**$^+$ results in a sequence **H** admitting a trivial specification $\mathbf{H}_c$, the result of this application of Rule **Б**$^+$ will be considered to be $\mathbf{H}_c$ (not **H**).

4. Use of various properties of favorable sequences (including some not mentioned in this section) leads to the fact that one verifies the favorability of the empty sequence more conveniently, generally speaking, with the help of an object different from a foundation of favorability of this sequence. For example, use of factorability gives us a branching process, and an object "justifying favorability" is more conveniently represented "nonlinearly". In this respect, let us consider "foundations", basing ourselves on Lemma 8.

Let us assume that there is an application of Rule **Б**, with the notation of p. 18, and assume that Ξ_i denotes the set of numbers mentioned in the characteristic with respect to which $\mathbf{H}_i$ is a subsequence of $\mathbf{H}'_i$ $(i = 1, \cdots, \nu)$. We call this a *decomposing application* (or *dc-application*) if $\Xi_1, \cdots, \Xi_\nu$ pairwise intersect only in the number $h + 1$. We say that an application of Rule **Б**$^+$ is a *dc-application* if the arguments of this application have no trivial specifications and the result coincides with the result of some dc-application of **Б** to the same arguments or is obtained from it by a trivial specification. An application of Rule $\mathbf{A}^+$ is called a *dc-application* if its result is not a nontrivial specification of a +-closed sequence. For example, an application of $\mathbf{A}^+$ giving the sequence $[1; x^1 \approx y^1]$ for the sequent

$$\rightarrow \exists xy\,(Pxy \vee \neg Pyx)$$

is not a dc-application, since this sequence is obtained by a contraction operation from the +-closed sequence $[1, 1; x^1 \approx y^2, y^1 \approx x^2]$. We remark that a dc-application is best suited for presentation in operator form.

Let us introduce a one-premise rule **Sp**: as a result of applying **Sp** to **H** we take any of its specifications $\mathbf{H}_c$. Any sequence obtained by $\mathbf{A}^+$ or $\mathbf{Б}^+$ can be obtained by applying **Sp** to the result of a suitable dc-application of $\mathbf{A}^+$ or $\mathbf{Б}^+$. We shall say that a list of sequences is a *dc-foundation* (relative to $\mathbf{H}_1, \cdots, \mathbf{H}_f$) of its last term if every sequence **H** in the list which is not the result of a dc-application of $\mathbf{A}^+$ (and the sequence $\mathbf{H}_1, \cdots, \mathbf{H}_f$) is obtained from the preceding ones either by **Sp** or by a dc-application of $\mathbf{Б}^+$. (In addition, one can deduce the following restriction: in a dc-foundation there are no two applications of **Б** with the same list of arguments.) From Lemmas 8 and 6 one obtains the possibility of a restriction to dc-foundations alone in the process of establishing deducibility. dc-foundations are more useful to us in practice than the usual foundations of favorability. Therefore in the sequel many results will be proved for dc-foundations, even though corresponding analogues also hold for foundations.

§3. П-tactics

1. To decrease the verifications arising in the process of establishing deducibility and to increase the practical effectiveness of algorithms based upon the inverse method, the possibility of introducing various restrictions on the process of constructing favorable sequences has great value. By a *tactic* we mean a system, formulated in an arbitrary way, of restrictions on the choice of arguments for sequential application of Rule **Б**, on the form of permitted applications of this rule, on the consideration of closed sequences, etc. We encountered the simplest examples of such restrictions in the preceding section; thus the permission to use only Rules $\mathbf{A}^+$ and $\mathbf{Б}^+$ and the prohibition on the use, as arguments for Rule $\mathbf{Б}^+$, of sequences, proper sequences of which have already been found to be favorable, form a certain tactic for establishing deducibility. A tactic is said to be *complete* if, for any derivable formula, one can construct a foundation for favorability of the empty sequence which satisfies all the restrictions imposed by this tactic.

A significant difficulty in verification is connected with the following fact: for any favorable sequence we know that a great number of different variants of the sequence (permutations, contractions, etc.) are favorable, but we cannot predict which of these variants are necessary for obtaining the empty sequence. In this section we shall study the possibility of decreasing the number of necessary variants of sequences (in terms of dc-foundations, the possibility of restricting the application of **Sp**).

Let us first consider *fusions*, that is, results of applications of **Sp** having smaller length than the premise.[8] Let **H** be a sequence and $\mathbf{H}_c$ one of its fusions. It is always easy to find sequences $\mathbf{H}_1, \cdots, \mathbf{H}_f$ such that one can obtain the empty sequences by an application to $\mathbf{H}_1, \cdots, \mathbf{H}_f$ and **H** of Rule $\mathbf{Б}^+$, but one cannot arrive at the empty sequence from the sequences $\mathbf{H}_1, \cdots, \mathbf{H}_f$ and **H** by any applications of **Б** and **Sp**. Therefore, to be able to obtain favorable sequences **H** and $\mathbf{H}_c$, we must keep at least **H** for the following process of constructing favorable sequences. The need for using $\mathbf{H}_c$ is not so important: as a rule, the favorability of an empty sequence can be justified by using only "decomposed variants" of sequences. Nevertheless the complete rejection of fusions is impossible. In fact, consider the formula

$$\exists xy\, ((Pxy \vee \neg Pyx) \,\&\, (Qxy \vee \neg Qyx)).$$

A dc-application of $\mathbf{A}^+$ gives us the sequences $[1, 1; x^1 \approx y^2, x^2 \approx y^1]$ and $[2, 2; x^1 \approx y^2, x^2 \approx y^1]$, to which we apply $\mathbf{Б}^+$, obtaining the two sequences

$$[1,\ 2;\ x^1 \approx x^2,\ y^1 \approx y^2], \quad [2,\ 1;\ x^1 \approx x^2,\ y^1 \approx y^2].$$

No application of $\mathbf{Б}^+$ to the indicated sequences gives us a new sequence; that is, if we do not allow **Sp**, it is impossible to obtain an empty sequence. On the other hand, if we permit the use of the closed sequence $[1; x^1 \approx y^1]$ (that is, if we permit one application of **Sp** or of $\mathbf{A}^+$, which is not a dc-application), we easily justify the favorability of the empty sequence.

The necessity in the cited example of using fusions is, as is easily seen, connected with the following fact: each of the components of the sequence $[1, 1; x^1 \approx y^2, x^2 \approx y^1]$ is "not simpler" than the single component of the sequence $[1, x^1 \approx y^1]$. In this connection, it is convenient to consider the concept of an *aggravating fusion* (and its opposite, a *facilitating dispersion*). The idea of these concepts consists in the fact that the structure of every component undergoing contraction in the sequence being fused must be simpler than the structure of the component of the fused sequence which results from these contractions. The new terminology can be made precise in various ways; it is essential only that every sequence have a finite number of facilitating dispersions. In this connection, it is usefult to take into account the following: the greater the number of fusions that turn out to be aggravating, the more often we are able to avoid application of **Sp**.

(8) A fusion is any specification which is not a permutation.

It is advisable to make the indicated concepts precise in the following way. Returning to the definition of *contraction operation* (p. 21), we say that the substitution, mentioned in that definition, of the term T_u for the variable u *aggravates* the bundle if one of the following two conditions is fulfilled: 1) T_u is not a variable (but is, possibly, a zero-place symbol), and 2) u occurs in S only once and the variable T_u occurs in F-terms of S' not less than twice. A contraction operation is said to be *aggravating* if one can find u_1 and u_2 (not necessarily different) in S such that the variable u_1 (respectively u_2) occurs in S in a relation for an atom with upper index $h-1$ (respectively, h) and the substitutions for each of these variables aggravate the bundle.(9) A specification $\mathbf{H}_c$ of a sequence $\mathbf{H}$ is said to be an *aggravating fusion* (and $\mathbf{H}$ a facilitating dispersion of $\mathbf{H}_c$) if $\mathbf{H}_c$ is a fusion of $\mathbf{H}$ and all the contraction operations applied in the construction of $\mathbf{H}_c$ from $\mathbf{H}$ are aggravating. Equality of sequences, trivial contractions and permutations are special cases of nonaggravating specifications. If we introduce the notion of complexity of F-terms of a bundle, considering the complexity to be increased under any substitution aggravating the bundle (see footnote 9), then it is natural to call the total complexity of F-terms in relations for atoms with given upper index k the complexity of the kth component of the sequence. By induction with respect to the complexity of decomposed components it is easy to prove that any sequence has only finitely many facilitating dispersions.

A sequence $\mathbf{H}'$ is said to be a generalized subsequence of a sequence $\mathbf{H}$ if $\mathbf{H}'$ or one of its aggravating specifications is a subsequence of $\mathbf{H}$. We shall say that a dc-foundation for favorability of some sequence satisfies the *ss-tactic* (*tactic of subsequences*) if to the right of any sequence $\mathbf{H}$ in this foundation there does not occur an application of **Б** to an improper supersequence of any generalized subsequences of the sequence $\mathbf{H}$ (in particular, in such a foundation fusion of any sequence $\mathbf{H}$ can be used to the right of $\mathbf{H}$ only if it is not aggravating).

We shall say that a dc-foundation of favorability of some sequence satisfies the *dis-tactic* (*dispersion tactic*) if **Sp** is applicable in such a way that the result is a nonaggravating specification of the premise. As we shall see below, the combination of the ss-tactic and the dis-tactic is complete.(10)

---

(9) It is not difficult to understand in what sense contracted components are simpler than the result of contraction. This definition would be especially intuitive for a formulation of bundles in a language with nontotal functional symbols (see Remark 1). In fact, a variable having a unique occurrence in a bundle would be eliminated (with incorporation of the corresponding argument in the functional symbol; see the incorporation of the variable t in the example of Remark 1); that is, terms in relations of an aggravated fusion will simply be longer than the corresponding terms in the contracted components.

(10) The dis-tactic is a modification of a tactic proposed by V. P. Orevkov. Discussions with Orevkov of problems connected with the dis-tactic had great value for clarifying the difficulties that would arise and for making concepts and results precise.

2. Let us now study the possibility of eliminating permutations of favorable sequences. By an ordering of components in F-sequences (briefly, an ordering) we mean any decidable predicate Π which is defined on all F-sequences and, for any sequence, is true for at least one of its permutations; a sequence satisfying this predicate is said to be Π-ordered. We shall say that a foundation (dc-foundation) of favorability of some sequence satisfies the Π-*tactic* if all sequences in this foundation are Π-ordered. A Π-tactic the more strongly restricts the verifications in the process of establishing deducibility the smaller the fraction of Π-ordered sequences.

Let us show that not for every Π is the Π-tactic complete. We consider the sequent

$$\to \exists xy\,(Pxy \,\&\, Pyx),\ \exists \bar{x}\bar{y}\,(\neg P\bar{x}\bar{y} \,\&\, \neg P\bar{y}\bar{x})$$

with the decomposition $Pxy, Pyx, \neg P\bar{x}\bar{y}, \neg P\bar{y}\bar{x}$. Assume an ordering Π such that, of the eight +-closed sequences of this formula, the following are Π-ordered:

$$[1,\ 3;\ x^1 \approx \bar{x}^2,\ y^1 \approx \bar{y}^2],\quad [4,\ 1;\ \bar{x}^1 \approx y^2,\ \bar{y}^1 \approx x^2],\quad [3,\ 2;\ \bar{x}^1 \approx y^2,\ \bar{y}^1 \approx x^2],$$
$$[2,\ 4;\ x^1 \approx \bar{x}^2,\ y^1 \approx \bar{y}^2].$$

Applying $Б^+$ to the first and last sequences, we obtain $[1, 2; x^1 \approx x^2, y^1 \approx y^2]$; assume that this sequence is Π-ordered but its permutation is not. Analogously, we can obtain a Π-ordered sequence $[3, 4; x^1 \approx x^2, y^1 \approx y^2]$ with a non-Π-ordered permutation. It is evident that, starting from the six sequences that we have obtained, we cannot obtain a new Π-ordered sequence by an application of $Б^+$. Among other things, even if permutations of the last two sequences were Π-ordered, we still could not obtain an empty sequence. Thus, if the ordering Π is as described, the Π-tactic will not enable us to establish deducibility of the given formula. The reason for this "circularity" of the process of obtaining favorable sequences is that the ordering "gives contradictory instructions" as to preferability of the form of the last (the outer ones for an application of **Б**) components of Π-ordered sequences. This "inconsistency" shows itself in the following way: excision from the sequence

$$[1,\ 3,\ 2,\ 4;\ x^1 \approx \bar{x}^2 \approx y^3 \approx \bar{y}^4,\ y^1 \approx \bar{y}^2 \approx x^3 \approx \bar{x}^4]$$

with respect to characteristic 1, 2, and 2, 3, and 3, 4 leads to Π-ordered sequences, but excision with respect to characteristic 1, 4 yields a sequence for which only a permutation is Π-ordered. It turns out that, if we prohibit for Π the occurrence of such an effect, we arrive at a complete Π-tactic.

A subsequence of a sequence **H** with respect to some characteristic is said to be *maximal* if it is obtained from **H** by excision with respect to the characteristic. We shall call an ordering Π *acyclic* (*weakly acyclic*) if, for any Π-ordered sequence, any subsequence (respectively, maximal subsequence) is Π-ordered.

THEOREM 2. *For any derivable formula F and any acyclic ordering Π, one can construct a dc-foundation of favorability of the empty sequence, simultaneously satisfying the Π-tactic, dis-tactic, and ss-tactic.*

For the proof one makes use of the following.

LEMMA 9. *For any derivable formula F and any weakly acyclic ordering Π, one can find supersequences of +-closed sequences (having the same length as the corresponding +-closed sequences), from which, by dc-applications of Rule $Б^+$, one can obtain a foundation of favorability of the empty sequence, satisfying the Π-tactic and not containing an application of Б to sequences whose maximal subsequences are already in the foundation.*

3. The proof of the lemma will be carried out in subsection 4, and the proof of the theorem in subsection 6, but before that we shall introduce some definitions. We call a sequence $[C; S]$ *constant* if, in S, 1) all possible (F, C)-atoms are mentioned, and 2) none of the F-terms used contains variables (a nonempty sequence can be constant only if F contains zero-place functional symbols, and therefore we shall assume that in F there occurs at least one such symbol, if only a fictitious one). We say that a foundation of favorability is *constant* if it contains only constant sequences. Below, a foundation of favorability will often be written in the form of a tree, consisting of applications of Rule Б; such trees, composed of constant sequences, will also be called *constant.*

The basic property of constant sequences is the uniqueness of the formula corresponding to every such sequence and the uniqueness of the form of the formula, the existence of which in a branch guarantees that in this branch there occurs a given constant sequence (for arbitrary sequences such formulas are determined up to substitutions for variables). We shall also use other properties of constant sequences; a constant subsequence of a sequence is a maximal subsequence; a constant sequence is factorable into one-termed factors; all one-termed factors, except the last,(11) of a constant sequence which is an argument of an application of Rule Б are factors of the result of this application, etc.

In studying tactics it is convenient to consider first the corresponding tactic on constant foundations, and only then to proceed to foundations of sequences of a general form. This first stage corresponds, in a certain sense, to a "reduction to the propositional calculus"; that is, to consideration of a certain Herbrand expansion. Only at the second stage does the specific character of the predicate calculus reveal itself and one obtains results whose basic content disappears when they are formulated in terms of propositional formulas. In accordance with what has been said, in the proof of

(11) This last factor will also be called the outer component of the application of Б (compare footnote 4).

Lemma 9 the required foundation constructed by us is constant, and only in going over to the proof of Theorem 2 do we need sequences of a general form. We note that the ss-tactic and the dis-tactic are essentially "quantificational" and become trivial on constant sequences. Therefore the important point in Theorem 2 is, in fact, the necessity of using in the general case nonconstant sequences (arising from the requirement that the ss-tactic uses only +-closed sequences). Lemma 9 could be strengthened somewhat at the cost of a transition to the "level of the predicate calculus"; the proof would then have to be developed according to the scheme in [9], §3. The basic meaning of such a strengthening (and of the increased complexity of the construction) consists of the possibility of obtaining an analogue of the theorem for weakly acyclic orderings. In fact, such an analogue holds, but one cannot use the ss-tactic for this purpose. It is not difficult to construct a derivable formula F and a weakly acyclic ordering Π for which it is impossible to obtain a foundation of favorability of the empty sequence satisfying the Π-tactic, without having used a sequence, a permutation of which is a supersequence of a +-closed sequence. In the view of the author, this fact makes it inappropriate to use the Π-tactic with weakly acyclic (but not acyclic) Π.

4. Proof of lemma 9. We take an arbitrary foundation of favorability of the empty F-sequence, write it in the form of a tree and convert this tree into a constant one, replacing closed sequences at the very top by corresponding constant supersequences (preserving the length of sequences). We do this in order that all applications of Rule **Б** in this tree will be applications of **Б**$^+$; then they will be dc-applications of **Б**$^+$ (since all the contractions used will be trivial). We denote this tree by $\mathfrak{D}_0$, and by an *element* of this tree we mean any one-termed constant sequence which is a factor of at least one of the sequences occurring in $\mathfrak{D}_0$. Let Ψ be a Π-ordered constant sequence whose factors are all elements of $\mathfrak{D}_0$ (each occurs in Ψ once). Until the end of this subsection, in speaking about trees we shall have in mind constant trees of dc-applications of Rule **Б**$^+$, at the top of which stand closed sequences and all elements of which occur in Ψ. In speaking about Π-ordered sequences, we shall have in mind certain special cases: sequences obtained by excision from Ψ.

Let $\mathfrak{D}$ be a tree with lowest sequence $\mathbf{H}_{\mathfrak{D}}$. By the *improper part* of this tree we mean the result of eliminating from $\mathfrak{D}$ all Π-ordered sequences which were obtained by using only Π-ordered sequences. Let $\Xi_{\mathfrak{D}}$ be a list of the elements of this tree which occur in the improper part but do not occur in $\mathbf{H}_{\mathfrak{D}}$. Assume that $\mathbf{H}_{\mathfrak{D}}$ is Π-ordered and that all of its elements occur in Ψ to the left of all elements of $\Xi_{\mathfrak{D}}$. By induction with respect to the length of the list $\Xi_{\mathfrak{D}}$ we prove the possibility of constructing a tree of Π-ordered sequences ending with a maximal subsequence of $\mathbf{H}_{\mathfrak{D}}$. The base of the induction is obviously fulfilled, since it reduces to the case where there is no improper part in the tree.

Let us proceed with the induction step. Assume that ξ is an element of $\Xi_{\mathfrak{D}}$ occurring in Ψ to the left of the other elements of $\Xi_{\mathfrak{D}}$. Let us consider all occurrences of ξ in the improper part in which ξ turns out to be the outer component of an application of Б. Now we do not carry out these applications and we alter, correspondingly, the part of the tree lying below the occurrences being considered (ξ enters into some sequences as a factor; in the remainder, this part only will be simplified). Above the applications of Б being considered, we move ξ in all sequences of the improper part to the left of all other elements of $\Xi_{\mathfrak{D}}$; in П-ordered sequences, which are boundaries of the improper and proper parts, ξ stands exactly between factors of $\mathbf{H}_{\mathfrak{D}}$ and the remaining factors of $\Xi_{\mathfrak{D}}$.(12) Therefore the proper part of the tree does not change and we obtain a tree $\mathfrak{D}'$ whose lowest sequence $\mathbf{H}_{\mathfrak{D}'}$ does not contain a factor different from those occurring in $\mathbf{H}_{\mathfrak{D}}$ and ξ. The corresponding list $\Xi_{\mathfrak{D}'}$ is shorter than the list $\Xi_{\mathfrak{D}}$ (by at least the element ξ). Using the inductive hypothesis, we construct a tree $\mathfrak{D}''$ without improper part (its lowest sequence $\mathbf{H}_{\mathfrak{D}''}$ is a maximal subsequence of $\mathbf{H}_{\mathfrak{D}'}$). We return to $\mathfrak{D}$. Let $\mathbf{H}^{\xi}$ be one of the lowest sequences of this tree, a factor of which is ξ. We replace $\mathbf{H}^{\xi}$ and everything above it by the tree $\mathfrak{D}''$, we change correspondingly the part of the tree lying below it (moreover, to certain sequences are added factors of $\mathbf{H}_{\mathfrak{D}'}$); for the rest, the tree becomes simpler. Carrying out this change for any such sequences $\mathbf{H}^{\xi}$, we obtain a tree $\mathfrak{D}'''$. Since the part of this tree equal to $\mathfrak{D}''$ enters now into the proper part, and ξ vanishes with each application of Б to $\mathbf{H}^{\xi}$ (and does not appear again, since the sequences $\mathbf{H}^{\xi}$ were chosen to be the lowest), it follows that $\Xi_{\mathfrak{D}'''}$ is shorter than $\Xi_{\mathfrak{D}}$. A second application of the inductive hypothesis yields the required result.

Since the initial constant tree satisfies all the requirements which were assumed in carrying out the induction, we can construct a constant foundation of the empty sequence, consisting of П-ordered sequences, beginning with suitable supersequences of +-closed sequences and containing only dc-applications of Б^+. In this foundation it only remains to substitute suitable subsequences for the arguments of all applications of Б to sequences whose maximal subsequences have already been obtained in the foundation.

By the fact that П is weakly acyclic, this does not disturb the П-ordering. This proves the lemma.

5. Let us impose an additional requirement on the sequence Ψ of the preceding subsection. We shall say that the sequence $[\alpha_1, \cdots, \alpha_{k-1}, \alpha_k, \alpha_k, \alpha_{k+1}, \cdots, \alpha_h; S']$ is obtained from (6) by duplication of the kth component if S' is obtained from S by replacing every relation of the form $u^l \approx T$ (for $l > k$) by the relation $u^{l+1} \approx T$, and by adding, for every relation $u^k \approx T$, the relation $u^{k+1} \approx T$. We denote by Ψ_N the sequence obtained from Ψ by N-fold duplication of every component.(13)

(12) Neither ξ nor elements of $\mathbf{H}_{\mathfrak{D}}$ and $\Xi_{\mathfrak{D}}$ are obliged to occur in these boundary sequences. On the other hand, the last of its components can be an element, which generally is not in the improper part.

(13) The length of Ψ_N exceeds that of Ψ by a factor of $N + 1$.

We choose N large enough so that every facilitating dispersion of any subsequence of Ψ will be a permutation of some subsequence of Ψ_N (by the finiteness of the number of facilitating dispersions, such an N exists). Henceforth we shall assume that Ψ_N is Π-ordered.

Let us show that this requirement does not restrict generality; that is, that for $\mathfrak{D}_0$ one can always find a Π-ordered Ψ with Π-ordered Ψ_N. Assume that Ψ' is an arbitrary sequence containing all elements of the tree $\mathfrak{D}_0$. We duplicate each of its components Nh times, where h is the length of Ψ'. We find a Π-ordered permutation Ψ'' of the resulting sequence. We eliminate from the resulting sequence as many first components as is necessary so that in the eliminated part (whose length turns out not greater than $Nh + 1$) some factor occurs $N + 1$ times. We take this factor as the first factor of the sequence Ψ. Carrying out the described operation h times (later factors of Ψ must be chosen to be different from those already found), we construct the required sequence Ψ, for which Ψ_N (and Ψ itself) is obtained from Ψ'' by excision and, therefore, is Π-ordered.

In what follows, in the proof of Theorem 2 we shall assume Π to be acyclic and we shall consider only various subsequences of the sequence Ψ_N (which will be called Ψ_N's). From this it will follow that all sequences used are Π-ordered and that the number of all foundations composed of these sequences (and not containing repetitions) is finite.

6. PROOF OF THEOREM 2. Assume that Π is acyclic, that Ψ and Ψ_N are as in subsection 5, and that $\mathfrak{D}_0$ is a dc-foundation constructed from Ψ_N's in accordance with Lemma 9. Replacing every top sequence in this foundation by a corresponding +-closed sequence, we arrive at a subsequence (by applying $\mathbf{Б}^+$); in cases where in $\mathfrak{D}_0$ there was applied a trivial contraction which cannot be carried out for subsequences, we take a dc-application of $\mathbf{Б}^+$ with a subsequent application of **Sp**. If the result of this application of **Sp** is not an aggravating fusion of the argument, then the dis-tactic does not require its elimination.[14] The argument of such an application of **Sp** cannot be a Ψ_N. Since the same argument is not needed for the following application of **Б**, one can take any of its Π-ordered permutations (the result of such an application is already required to be a Ψ_N). After this elimination, all the sequences used will be Ψ_N's; that is, this exclusion does not vitiate the Π-tactic nor change the finiteness of the number of foundations under consideration. Thus we have a dc-foundation satisfying the Π-tactic; introducing into it (if necessary) some more applications of rule **Sp**, we make it satisfy the ss-tactic also. We write it linearly and denote it by $\mathfrak{O}$.

---

(14) If such an application can be divided into two, a "nonaggravating" $\mathbf{Sp}_1$ and an aggravating $\mathbf{Sp}_2$, then of course it is only necessary to leave $\mathbf{Sp}_1$, while $\mathbf{Sp}_2$ is eliminated from the common foundations.

We shall move from left to right in $\mathfrak{D}$, transferring aggravating applications of **Sp** as far to the right as possible. We obtain $\mathfrak{D}'$, whose initial segment consists of dc-applications of $\mathbf{A}^+$ and $\mathbf{Б}^+$ and applications of **Sp** which are not eliminable (we call this segment "*regular*"); then there follows a chain of aggravating applications of **Sp**, generating an argument-sequence for the next chain of dc-applications of $\mathbf{Б}^+$. In this chain we split all applications into minimal ones; that is, reducing the length of the argument to 1. We move consecutively each of the resulting aggravating applications of **Sp** past the nearest dc-application.

Assume that by an aggravating fusion of the kth and $(k + 1)$th components of a sequence $\mathbf{H}$ one obtains $\mathbf{H}_c$, to which $\mathbf{Б}$ is applied with result $\mathbf{H}_0$ and other arguments Δ. We carry out a dc-application of $\mathbf{Б}^+$ to $\mathbf{H}$ and Δ; the resulting sequence is denoted $\mathbf{H}'_0$. If a dispersion does not affect the outer components of the application (that is, if the length of $\mathbf{H}$ exceeds $k + 1$), then a single application of **Sp** to $\mathbf{H}'_0$ yields $\mathbf{H}_0$ or one of its subsequences. If only the outer component of the application of $\mathbf{Б}$ is subject to dispersion, then $\mathbf{H}'_0$ contains, by comparison to $\mathbf{H}_0$, more "successors" of the kth component of sequence $\mathbf{H}$. Then it is necessary for us to simulate in the application to $\mathbf{H}'_0$ the same work that was carried out in $\mathfrak{D}'$ in the application to $\mathbf{H}_0$. This simulation must proceed as long as the mentioned "successor" does not leave at the right-hand end of its sequence (at the end of the simulated part the sequence $\mathbf{H}$ stood before, but now there stands the sequence $\mathbf{H}'$, obtained from $\mathbf{H}$ or one of its subsequences by writing the appropriate "successor" to the right of it). We take $\mathbf{H}'$ as argument of a dc-application of $\mathbf{Б}^+$ with other arguments Δ. To the result we apply the same simulated part and arrive, finally, at the sequence $\mathbf{H}_0$ (or one of its subsequences). We remark that the indicated transition to a subsequence of $\mathbf{H}_0$ arises from possible trivial contractions (trivial "insertions" in a "successor").

Thus we have succeeded in decreasing the number of eliminable applications of **Sp** before the application of $\mathbf{Б}$ being considered. To the right of the position under consideration there can occur repetitions of sequences and violations of the ss-tactic, which are easily eliminated. By a multiple application of the described transformations we are able to bring the dc-application of $\mathbf{Б}^+$ immediately to the proper part. As a result we obtain a dc-foundation of the empty sequence, satisfying the ss-tactic and Π-tactic with an expansion of the proper part. As we have already said, it is impossible to expand the proper part indefinitely; that is, the whole foundation finally becomes proper, satisfying the dis-tactic. The theorem is proved.

7. The simplest example of an acyclic ordering Π arises from any ordering π of the formulas of the decomposition (2). Π-ordered sequences are declared to be those for which every component α of the list precedes every component β whenever the αth formula of the decomposition precedes the βth formula in π. Assume that all disjunctions of (2) are one-termed and that $\bar{\pi}$ is an ordering under which all elementary formulas of (2) beginning with a negation sign come after atomic formulas.

The acyclic ordering Π which arises on the basis of $\bar{\pi}$ gives us a Π-tactic a special case of which is an analogue of the method of hyperresolution (for example, [3], p. 694) for the inverse method.(15) Similarly one obtains an analogue of the method of semantic P-deductions [3], etc.

More complicated acyclic orderings (based not only upon the list of the sequence but also on the bundle) permits a still stronger limitation on verifications and helps us to prove a further strengthening of the theorems about completeness of the hyperresolution tactic, of Theorem 8 of [3], etc. (cf. [4], [6], [7]). In this connection it is convenient to put on the left components of sequences, proper variables of which are set equal to simpler terms in the bundle. In fact, in the case of such an ordering, applications of Rule Б cut out more structurally complicated terms of the component (this improves the resulting sequences and decreases the steps in the choice of arguments for the application of Б). In particular, a convenient acyclic ordering arises on the basis of the following prescription: the kth component of sequence (6) is placed to the left of the lth component if there is in S a relation $u^l \approx T$ and, for every basic variable v of the α_kth formula of the decomposition, there occurs in S a relation $v^k \approx T'$, where T' is a proper subterm of T. This prescription by itself can serve as a basis for some acyclic ordering(16) and can be combined with other ordering principles. The occurrence in S of the kind of relations mentioned in this prescription is very characteristic of formulas in which, before Skolemization, are found positive quantifiers not containing negative quantifiers in their scope.

§4. Tactics combined with the support tactic

1. The connection described in [6] betwen the inverse method and the method of resolution opens up the possibility of a useful interpenetration of the two methods, of a transfer of the tactics of one to the other, etc. For the method of resolution the support tactic [2] is well known. This tactic limits to a small degree the verifications which arise; nevertheless it is of significant interest, since it permits us, in searching for a proof in axiomatic theories, to concentrate our search on the theorem being tested (not extracting, as far as is possible, those consequences of the axioms which are not necessary for a proof of the theorem). We shall formulate an analogue (which is at the same time somewhat of a generalization) of the tactic of supports, for the inverse method.

(15) For a complete ordering $\bar{\pi}$, it is natural to call this tactic a *strong tactic of hyperresolution.* A simple tactic of hyperresolution is obtained if $\bar{\pi}$ is a partial ordering which does not express any preference among atomic formulas of the decomposition and among elementary formulas beginning with a negation sign. The simple tactic of hyperresolution guarantees that, in the list of any resulting sequence (except closed ones), indices of nonatomic formulas of the decomposition will not be present.

(16) An ordering close to this was proposed for the deducibility determination algorithm by V. P. Orevkov [13].

Let M be some set of formulas of the decomposition (2). We say that a closed sequence of minimal length([17]) is *supported* by the set M if at least one of the elements of its list is an index of a formula of M. We shall say that a foundation (dc-foundation) of favorability satisfies the *support tactic* with support set M (sup M-tactic) if it does not contain any application of Rule **Б** all arguments of which are closed and not supported by M. Of the greatest importance is the choice of support sets connected with a selection of some of the formulas of sequent (1) (for example, formulas $F_1, \cdots, F_{n'}$, where $n' < n$) and with the disjunctions occurring in these formulas (in the example, the disjunctions $D_{1,1}, \cdots, D_{1,\delta_1}, \cdots, D_{n',1}, \cdots, D_{n',\delta_{n'}}$). In this case the sup M-tactic ensures the primary use, in the process of constructing favorable sequences, of the selected formulas (in the generation of each sequence one uses subformulas of the chosen components of the sequent). Let us consider first the sup M-tactic for constant trees.

We say that a constant tree of applications of $\mathbf{Б}^+$ and **Sp** is *regular* (compare [14]) if above each sequence **H** in this tree there does not occur an application of Rule **Б** one of whose outer components is a one-termed subsequence of the sequence **H**. It is evident that, discarding from a nonregular tree those applications of Rule **Б** which violate regularity (and possibly some underlying applications of **Б** which have become superfluous) and adding some permutations according to rule **Sp**, we obtain a foundation with the same lowest sequence of one of its subsequences. The result of writing a sequence $\mathbf{H}_1$ to the left of a sequence $\mathbf{H}_2$ shall be denoted $\mathbf{H}_1 \cdot \mathbf{H}_2$. If $\mathfrak{M}$ is a set of sequences, and a tree of applications of **Б** and **Sp** with lowest sequence **H** does not contain closed sequences not occurring in $\mathfrak{M}$, then any supersequence of the sequence **H** will be said to be *$\mathfrak{M}$-derivable.*

Let $\mathfrak{D}_0$ be a constant tree which is a foundation of favorability of the sequence **H** with respect to sequences $\mathbf{H}_1, \cdots, \mathbf{H}_f$. We write a complete list without repetitions of one-termed factors of sequences of this tree in the form $\mathbf{H}^1, \cdots, \mathbf{H}^{s+h}$, where $\mathbf{H}^{s+1}, \cdots, \mathbf{H}^{s+h}$ are factors of **H**. In this subsection we shall consider only sequences which are factorable into nonrepeating factors from the list $\mathbf{H}^1, \cdots, \mathbf{H}^{s+h}$. Let M be some set of formulas of the decomposition, and $\{M\}$ and $\{M^c\}$ two sets of closed sequences of minimal length which are, respectively, supported and not supported by the set M. We denote by M_{ix} the set obtained by writing **H** to the left of each sequence of $\{M\} \cup \{\mathbf{H}_1, \cdots, \mathbf{H}_f\}$, removing repetitions of components of the sequences, and throwing out supersequences of sequences of $\{M^c\}$. Let $\{\overline{M}\}$ be a subset of $\{M^c\}$ such that in each sequence of $\{M^c\} - \{\overline{M}\}$ there occurs some $\mathbf{H}^\alpha$ with $\mathbf{H} \cdot \mathbf{H}^\alpha \in M_{ix}$.

LEMMA 10. *Assume that* **H** *is not* $\{\overline{M}\}$*-derivable. For* $\mathfrak{D}_0$, *one can construct a tree of applications of* $\mathbf{Б}^+$ *and* **Sp** *which is a foundation of favorability of* **H**

([17]) This is a sequence for which any excision leads to a nonclosed sequence.

with respect to sequences of M_{ix}, *and in which* 1) *sequences of* M_{ix} *take part in the generation of every nonclosed sequence, and* 2) *supersequences of closed sequences do not arise in applications of* **Б**.

PROOF. We write the sequence **H** to the left of each sequence of $\mathfrak{D}_0$ (removing repetitions of components of sequences), we convert the resulting tree into a regular one, we substitute the sequence $\mathbf{H}\cdot\mathbf{H}^{\alpha}$, where $\mathbf{H}\cdot\mathbf{H}^{\alpha}\in M_{\mathrm{ix}}$, for any of its supersequences, and we throw out all superfluous applications of **Б** and **Sp** (in particular, those lying above closed sequences). We denote the result by $\mathfrak{D}_1$. In the course of the induction carried out below, we shall use an extension of the sequence **H**, an expansion of M_{ix} and a corresponding contraction of the set $\overline{M}$ (preserving the relations described above among **H**, M_{ix}, and $\overline{M}$).

We shall call an occurrence of a sequence $\mathbf{H}^{\alpha}$ in the tree $\mathfrak{D}_1$ (as a factor of some sequence) "*good*" if it is an occurrence in a closed sequence or in a sequence of M_{ix} and it is an outer component of an application of **Б**. We carry out an induction with respect to the number t of sequences of $\mathbf{H}^1,\cdots,\mathbf{H}^s$ having "bad" occurrences in $\mathfrak{D}_1$. If $t=0$, then $\mathfrak{D}_1$ consists of one application of **Б** and the assertion of the lemma is obvious. Assume $t>0$. Consider the following cases.

Case 1. One can find a sequence $\mathbf{H}^{\alpha}$ having a "bad" occurrence and such that $\mathbf{H}\cdot\mathbf{H}^{\alpha}$ is not $\{\overline{M}\}$-derivable. Throwing out of $\mathfrak{D}_1$ applications of **Б** with outer component $\mathbf{H}^{\alpha}$ (and all applications of **Б** and **Sp** which become superfluous thereby), we obtain a tree which is a foundation of favorability of the sequence $\mathbf{H}\cdot\mathbf{H}^{\alpha}$. We transform it in the same way as in the transition from $\mathfrak{D}_0$ to $\mathfrak{D}_1$ (the role of **H** is played throughout by the sequence $\mathbf{H}\cdot\mathbf{H}^{\alpha}$, and M_{ix} is modified accordingly). For the transformed tree, we find ourselves with the conditions of applicability of the lemma. Moreover, the induction parameter for the new tree is lower, since the new "bad" occurrences arise only for factors of $\mathbf{H}^{\alpha}$, which do not enter now into the calculation of the number t. Therefore one can construct a tree $\mathfrak{D}^{\alpha}$, as described in the lemma, with $\mathbf{H}\cdot\mathbf{H}^{\alpha}$ in the role of **H**. Returning now to $\mathfrak{D}_1$, we extend M_{ix} by adding the sequence $\mathbf{H}\cdot\mathbf{H}^{\alpha}$ (as a sequence with respect to which one constructs a foundation of favorability) and in $\mathfrak{D}_1$ we replace sequences containing $\mathbf{H}^{\alpha}$ by the sequence $\mathbf{H}\cdot\mathbf{H}^{\alpha}$. After eliminating superfluous factors and applications of **Б** (and **Sp**) which have arisen, we obtain a tree not containing "bad" occurrences of $\mathbf{H}^{\alpha}$. Using the inductive hypothesis again, we obtain a tree, as described in the lemma, but containing among its top sequences the sequence $\mathbf{H}\cdot\mathbf{H}^{\alpha}$. Constructing over all these sequences the tree $\mathfrak{D}^{\alpha}$ (from which, possibly, one must eliminate some occurrences of factors of $\mathbf{H}^{\alpha}$), we conclude the consideration of this case.

Case 2. For all α $(1\leqslant\alpha\leqslant s)$, if $\mathbf{H}\cdot\mathbf{H}^{\alpha}$ is not $\{\overline{M}\}$-derivable, then $\mathbf{H}^{\alpha}$ occurs in $\mathfrak{D}_1$ only in a part of the sequence $\mathbf{H}\cdot\mathbf{H}^{\alpha}$ from M_{ix}. Using this M_{ix}, we contract $\{\overline{M}\}$ to the maximal possible extent. We again find ourselves with the conditions of applicability of the lemma; moreover, some new sequences of the form $\mathbf{H} \cdot \mathbf{H}^\beta$ can turn out to be non-$\{\overline{M}\}$-derivable. Then Case 1 can arise again, and the possibility of a new extension of M_{ix} and a new contraction of $\{\overline{M}\}$. Therefore we shall assume (on the basis of several alterations of M_{ix} and $\{\overline{M}\}$) that any factor $\mathbf{H}^\beta$ of any sequence of $\{\overline{M}\}$ $(1 \leqslant \beta \leqslant s)$ does not give a sequence $\mathbf{H} \cdot \mathbf{H}^\beta$ of M_{ix}. Let us prove that $\mathbf{H}^\beta$ does not have "bad" occurrences. We assume the contrary. Then $\mathbf{H} \cdot \mathbf{H}^\beta$ is not closed and we have a derivation in which only those factors $\mathbf{H}^\gamma$ $(1 \leqslant \gamma \leqslant s)$ are referred to for which $\mathbf{H} \cdot \mathbf{H}^\gamma$ is $\{\overline{M}\}$-derivable. Replacing in this derivation every closed sequence containing $\mathbf{H}^\beta$ and $\mathbf{H}^\gamma$ by the sequence $\mathbf{H} \cdot \mathbf{H}^\gamma$ (with a corresponding $\{\overline{M}\}$-derivation), we obtain an $\{\overline{M}\}$-derivation of $\mathbf{H}$, which contradicts our hypothesis. Thus all $\mathbf{H}^\beta$ have in the foundation only "good" occurrences; that is, there are no "bad" occurrences at all in the foundation. This proves the lemma (to obtain the desired foundation, one may need to build on top of sequences from the extensions of M_{ix} that are used; we are able to do this, since M_{ix} was extended only in cases where it was possible to construct a corresponding foundation).

2. COROLLARY. *For any favorable sequence which is not derivable without a supporting set* M *of sequences, one can find a generalized subsequence having a dc-foundation which satisfies the* sup *M-, ss-, and dis-tactics.*

In fact, with respect to constant sequences the content of the corollary immediately follows from Lemma 10 (one only requires a conversion of the foundation constructed in that lemma to closed sequences of maximal length, with deletion of resulting "redundancies"). In the conversion to nonconstant sequences it is necessary to be careful about the removal of generalized supersequences (with admission of arbitrary applications of **Sp**) and then about the rejection of applications of **Sp** prohibited by the dis-tactic (by the method of extension of the proper segment of the foundation; cf. §3.6). By virtue of the first of these steps, we note that the absence of supersequences of closed sequences among the results of the application of **Б** in the tree of Lemma 10 guarantees the following: upon replacing some nonclosed sequence by its subsequence, obtained earlier, this subsequence also is the result of an application of **Б**; that is, this replacement does not violate the sup M-tactic. In the second step, upon replacing a nonclosed sequence $\mathbf{H}$ by its essential dispersion $\mathbf{H}'$, a similar situation arises: $\mathbf{H}'$ is a result of an application of **Б**, and consequently is generated in accordance with the sup M-tactic. If $\mathbf{H}$ is obtained by $\mathbf{A}^+$, then the result $\mathbf{H}'$ of the corresponding dc-application of $\mathbf{A}^+$ will be supported (or not supported) by the set M in the same way that $\mathbf{H}$ is.

As a result we arrive at a proof of the following theorem (a support set M is said to be *necessary* if the derivability of the formula F cannot be established without sequences supported by M).

THEOREM 3. *For any derivable sequent* (1) *and any necessary support set* M, *one can construct a dc-foundation of favorability of an empty sequence, satisfying the* sup M-, *dis-*, *and ss-tactics.*

3. The addition of a tactic T' to a tactic T will be said to be *conservative* if every sequence derivable in accordance with T turns out to be derivable in accordance with both tactics. Completeness of a conservative addition is obviously equivalent to the completeness of the original tactic.

In the notation of p. 18, we call an application of Rule $\mathbf{Б}^+$ an *eng-application* if one can choose $\mathbf{H}_i$ $(1 \leqslant i \leqslant \nu)$ ($\mathbf{H}_i$ is called the *principal argument* of the application) such that the characteristic with respect to which $\mathbf{H}_i$ is a subsequence of $\mathbf{H}'_i$ includes the characteristic with respect to which $\mathbf{H}_j$ is a subsequence of $\mathbf{H}'_j$ for any nonclosed $\mathbf{H}_j$ $(1 \leqslant j \leqslant \nu)$. In other words, in the "nonexcised part" of the principal argument there must be "inserted" the nonexcised parts of all remaining nonclosed arguments of the application. We shall say that a foundation of favorability of some sequence satisfies the *engulfing tactic* (*eng-tactic*) if all applications of $\mathbf{Б}$ in this foundation are eng-applications.

THEOREM 4. *For any support set* M, *addition of the eng-tactic to the* sup M-*tactic is conservative* (*in particular, for any necessary* M *the combination of the eng- and* sup M-*tactics is complete*).

Obviously it suffices to prove the theorem for constant trees. Assume we have a constant tree $\mathfrak{D}$ which is a foundation of favorability of a sequence $\mathbf{H}$ with respect to sequences which have been obtained from a sequence $\widetilde{\mathbf{H}}$ by writing to its right one-termed sequences $\widetilde{\mathbf{H}}^1, \cdots, \widetilde{\mathbf{H}}^f$, and ending with an application of Rule $\mathbf{Б}$ of the form

$$\frac{\widetilde{\mathbf{H}} \cdot \widetilde{\mathbf{H}}^1, \ldots, \widetilde{\mathbf{H}} \cdot \widetilde{\mathbf{H}}^f,\ \mathbf{H}_1 \cdot \mathbf{H}^1, \ldots, \mathbf{H}_\mu \cdot \mathbf{H}^\mu}{\mathbf{H}},$$

where $\widetilde{\mathbf{H}}^1, \cdots, \widetilde{\mathbf{H}}^f, \mathbf{H}^1, \cdots, \mathbf{H}^\mu$ is a complete list of the outer components of the application. Assume that for every i $(1 \leqslant i \leqslant \mu)$ the tree $\mathfrak{D}_i$ leading to the sequence $\mathbf{H}_i \cdot \mathbf{H}^i$ satisfies the eng-tactic and begins with minimal closed sequences; that is, $\widetilde{\mathbf{H}} \cdot \widetilde{\mathbf{H}}^1, \cdots, \widetilde{\mathbf{H}} \cdot \widetilde{\mathbf{H}}^f$ are not used in $\mathfrak{D}_i$. By induction on the number of occurrences in $\mathfrak{D}$ of sequences which are results of applications of $\mathbf{Б}$, we show that we can construct a foundation of favorability of $\mathbf{H}$ with respect to those sequences, satisfying the eng-tactic. If $l = 1$, then the tree $\mathfrak{D}$ is what is required. Assume $l > 1$, and, for instance, $\mathfrak{D}_1$ ends with an application of the form

$$\frac{\mathbf{H}_{1,1} \cdot \bar{\mathbf{H}}^1, \ldots, \mathbf{H}_{1,\nu} \cdot \bar{\mathbf{H}}^\nu}{\mathbf{H}_1 \cdot \mathbf{H}^1}, \tag{9}$$

where $\bar{\mathbf{H}}^1, \cdots, \bar{\mathbf{H}}^\nu$ are the outer components of the application. Without restricting generality, we shall assume that $\mathbf{H}^1$ occurs in $\mathbf{H}_{1,1}, \cdots, \mathbf{H}_{1,\nu'}$ ($\nu' \leqslant \nu$) and does not occur in $\mathbf{H}_{1,\nu'+1}, \cdots, \mathbf{H}_{1,\nu}$. Transposing the factor $\mathbf{H}^1$ into the last place in $\mathbf{H}_{1,1} \cdot \bar{\mathbf{H}}^1$,([18]) and applying Rule $\mathbf{Б}$ to the resulting sequence and to the sequences $\mathbf{H}_2 \cdot \mathbf{H}^2, \cdots, \mathbf{H}_\mu \cdot \mathbf{H}^\mu, \widetilde{\mathbf{H}} \cdot \widetilde{\mathbf{H}}^1, \cdots, \widetilde{\mathbf{H}} \cdot \widetilde{\mathbf{H}}^f$, we obtain a tree $\mathfrak{D}'_1$ with smaller parameter l. Similarly we obtain trees $\mathfrak{D}'_2, \cdots, \mathfrak{D}'_{\nu'}$. By the inductive hypothesis, all these trees can be formed in the required way. Assume first that the principal argument of the application (9) contains $\mathbf{H}^1$. Then an application of Rule $\mathbf{Б}$ to the lower sequences of these trees and to the sequences $\mathbf{H}_{1,\nu'+1} \cdot \bar{\mathbf{H}}^{\nu'+1}, \cdots, \mathbf{H}_{1,\nu} \cdot \bar{\mathbf{H}}^\nu$ is an eng-application leading to $\mathbf{H}$. Assume now that $\mathbf{H}^1$ does not occur in the principal argument of the application (9). Then, for all i $(1 \leqslant i \leqslant \nu')$, $\mathbf{H}_{1,i} \cdot \bar{\mathbf{H}}^i$ is a closed two-termed sequence; that is, $\mathbf{H}_{1,i} = \mathbf{H}^1$. Now the lowest sequences of the trees $\mathfrak{D}'_1, \cdots, \mathfrak{D}'_{\nu'}$ do not afford us the possibility of immediately obtaining an eng-application of Rule $\mathbf{Б}$. However, now we can use the fact that the lowest sequences of these trees differ only in one-termed sequences $\bar{\mathbf{H}}^1, \cdots, \bar{\mathbf{H}}^{\nu'}$, and consequently there is a sequence $\bar{\mathbf{H}}$ such that the application leading to the sequence $\mathbf{H}$ can be written in the following way:

$$\frac{\bar{\mathbf{H}} \cdot \bar{\mathbf{H}}^1, \ldots, \bar{\mathbf{H}} \cdot \bar{\mathbf{H}}^{\nu'}, \mathbf{H}_{1,\nu'+1} \cdot \bar{\mathbf{H}}^{\nu'+1}, \ldots, \mathbf{H}_{1,\nu} \cdot \bar{\mathbf{H}}^{\nu}}{\mathbf{H}}.$$

Now, choosing $\bar{\mathbf{H}} \cdot \bar{\mathbf{H}}^1, \cdots, \bar{\mathbf{H}} \cdot \bar{\mathbf{H}}^{\nu'}$ as the sequences with respect to which we establish favorability, we have a decrease in the inductive parameter and the possibility of completing the induction (one obtains a foundation with respect to $\bar{\mathbf{H}} \cdot \bar{\mathbf{H}}^1, \cdots, \bar{\mathbf{H}} \cdot \bar{\mathbf{H}}^{\nu'}$, and the favorability of each of these sequences with respect to $\widetilde{\mathbf{H}} \cdot \widetilde{\mathbf{H}}^1, \cdots, \widetilde{\mathbf{H}} \cdot \widetilde{\mathbf{H}}^f$ has already been established).

From what has been proved the theorem immediately follows, if one takes into account that all the reconstructions that we have used do not change the form of the uppermost applications of $\mathbf{Б}$ (and consequently preserve the sup M-tactic).

REMARK 2. In view of the fact that the eng-tactic permits "making of exceptions" for closed sequences, application of this tactic is especially advantageous for those organizations of the process of establishing deducibility under which closed sequences are not considered on the same basis as other favorable sequences. This is the case, for example, in the organization of the search for an empty favorable sequence in the deducibility determination algorithm [13].

§5. General case. Decompositions with repetitions

1. The aim of the following sections is the extension of the formulations and results presented above to the case of formulas of arbitrary structure. The situation is

([18]) It is evident that a permutation of components of a favorable sequence (with a corresponding permutation of the overlying sequences) does not violate the eng-tactic.

that the requirement of a preliminary reduction of sequents to one or another standard form, generally speaking, reduces the practical effectiveness of the resulting algorithm for establishing deducibility. This unpleasant effect already arises if, in the construction of a standardized form, one uses propositional distributivity, and, even more so, if one extends the scope of quantifiers (see, for example, [15], [16]). Therefore, in what follows, we shall carry out only such standardizations which are not reflected in the structure of the formulas and their possible derivations. Such standardizations are given by Skolem's method of elimination of quantifiers (see, for example, [17]) and by the concept of strong equivalence ([1], §2.2).

As a result, without decreasing the quality of the machine algorithms obtained by the method based upon the proposed concretization, we can limit ourselves to initial sequents of the form

$$\rightarrow F_1, F_2, \ldots, F_n, \tag{10}$$

where $F_1, \cdots, F_n$ do not contain logical symbols other than $\vee$, &, $\neg$, and $\exists$, and contain the symbol $\neg$ only within elementary formulas (F will be used everywhere to denote the formula $F_1 \vee \cdots \vee F_n$). In fact, reducing an arbitrary derived sequent Σ to the form (10), we shall have the possibility of extensive mutual simulation of derivations of the sequents Σ and (10) (in this simulation one can use certain permutations and elimination of repeated parts). In addition, we can assume (10) closed (free variables are treated as zero-place functional symbols) and such that none of the formulas $F_1, \cdots, F_n$ has the symbols & and $\vee$ as principal connectives (the problem of deducibility of an arbitrary formula immediately reduces to the problem of deducibility of a finite number of multiple disjunctions, every component of which is elementary or begins with the quantifier $\exists$). Assume also that the proper variables of different occurrences of quantified parts in (10) are different.

In a number of cases in the sequel, we shall mark off a "characteristic" of a transformation by means of those functional symbols which arise from the elimination of positive quantifiers. It should be said that such an elimination can hamper certain theoretical applications of the method of the concretization being studied,([19]) but the technical difficulties of a presentation allowing positive quantifiers would be great and would have no essential significance.

([19]) By theoretical applications we have in mind, in particular, application of the inverse method to the theory of decidable classes and reduction classes (see [1], [18], [19]): often classes of formulas which can be easily formulated in terms of positive quantifiers receive an opaque and unnatural formulation in the case of elimination of such quantifiers. In addition, certain arguments of the present chapter can be carried over to other calculi, including some to which the Skolem method is not applicable (we are referring primarily to the constructive predicate calculus with functional symbols); this carry-over would be simpler if we formulated the concretization under the assumption that positive quantifiers are present.

REMARK 3. In one respect the introduction of one or another standardization would be objectionable: it makes it difficult to use identical subformulas occurring in the original sequent. For a sufficiently complicated formula $\mathfrak{A}$ the reduction of the sequent $\rightarrow \mathfrak{A} \supset \mathfrak{A}$ to the form (10) would already yield a sequent whose derivability is not obvious. However, the difficulty of comparing standardized formulas is incomparably lower than the difficulty of establishing derivability (moreover, it is not difficult to formulate a machine algorithm which, at the same time that it constructs a standardization, stores those subformulas which are the same in the original sequent). Therefore the only thing that we have to be concerned about in this connection is the possibility of establishing deducibility "at the level of large-scale units". This possibility will be provided (see Remark 5 at the end of §6 below).

2. Until now there was not the least value in considering multiple conjunctions and disjunctions as separate symbols or abbreviations. In what follows (for purely technical reasons) we shall consider them to be abbreviations. We add to the calculus $\mathbf{G}_0$ the rule

$$\frac{\rightarrow \Gamma,\ A,\ B,\ \Gamma'}{\rightarrow \Gamma,\ (A \vee B),\ \Gamma'}\ (\rightarrow \vee)$$

and we permit in the formulation of the rule $(\rightarrow \exists)$ the existence of quantifiers within the formula M (assuming only that M does not begin with $\exists$). The resulting calculus will be denoted $\mathbf{G}$.

3. The restriction, introduced in §1, on the form of sequents permits us to uniquely determine, for a formula of the decomposition, the place at which it occurs in (1). This makes the presentation of the method somewhat more convenient. For formulas of arbitrary structure it is impossible to require such uniqueness. We adopt the temporary convention that the place of a formula in the decomposition (in which repetitions are now possible) uniquely determines the place of the occurrence of this formula in the original sequent. Thus an arbitrary list

$$V_1, V_2, \ldots, V_\delta \ (\delta \geqslant 0) \tag{11}$$

of subformulas of the sequent (10) is said to be a decomposition of this sequent if there exists an algorithm which indicates, for every formula V of (11), its occurrence in (10) (in speaking about the occurrence of the formula V in (10), we have in mind the result of the operation of this algorithm).

We can carry over without alteration from §1 to the present case (the role of the list (2) is played by (11)) the definitions of basic variable, (F, C)-atom, F-term, (F, C)-bundle, system of values of atoms, F-sequence, excision with respect to a characteristic, extension and elementary extension of a sequence, the property of being a subsequence, equality of sequences, consistency of bundles, and, finally, formula corresponding to a sequence.

Let **B** be a branch of a proof search tree of sequent (10), and let X be an occurrence in it of some formula as a component of some sequent. Looking at **B** downwards from X, we easily find the occurrence in (10) of some subformula X_0 from which X arises, and we are able to write down the list of terms which were substituted for variables of X_0 in obtaining X. The resulting occurrence of X_0 is said to be a *preimage* of X, and the terms of the indicated list substituted for variables are called *X-values* of the corresponding variables. We shall say that sequence (6) occurs in **B** if one can find a list $X_1, \cdots, X_h$ of occurrences in **B** of certain formulas (as components of their sequents), the preimages of which are occurrences in (10) of the formulas $V_{\alpha_1}, \cdots, V_{\alpha_h}$, and the system of values of (F, C)-atoms consisting of the X_k-values $(k = 1, 2, \cdots, h)$ satisfies S. (The corresponding definition of § 1 is a special case of the one just presented.) One can define in a natural way the sequents containing (6) in **B** (now it can turn out that there are fewer than h of them). The list $X_1, \cdots, X_h$ is called a *realization* of (6) in **B**. Analogues of Lemmas 1 and 2 are easily proved.

Occurrences of any two different formulas W and W' of the decomposition in (10) either do not intersect or are such that one is contained in the other. We shall say that W *properly occurs* in W' if an occurrence X in (10) of the first formula is situated within an occurrence X' of the second formula, and no occurrence of a third formula of the decomposition contains X and is contained in X'. We extend this definition also to the case $W' = F$. We shall say that the kth component of sequence (6) is *subordinate* to the lth $(1 \leqslant k, l \leqslant h)$ if W_k properly occurs in W_l and for any basic variable u of the formula W_k (it will also be a basic variable of W_l) one can find in S relations $u^k \approx T$ and $u^l \approx T$.[20] We shall say that the kth component of sequence (6) is *subordinate to the zeroth* if there is a proper occurrence of W_k in F.

We shall define inductively the notion: "the collection $Y_1, \cdots, Y_\sigma$ of occurrences in the formula Y is an *obstruction* in Y" (compare [20]):

1) If Y is an elementary formula, then its only obstruction is the occurrence of Y in Y.

2) If $Y = \exists xA$ $(Y = A \vee B)$, then the only obstructions in Y are all the obstructions in the formula A (and all the obstructions in B) and the occurrence of Y in Y.

3) If $Y = A \mathbin{\&} B$, then any obstruction in A, combined with any obstruction in B, yields an obstruction in Y, and the only other obstruction is the occurrence of Y in Y.

Interchanging the positions of the symbols & and $\vee$ in this definition, we obtain a definition of a *path* in Y. We remark that it is impossible to eliminate any element from an obstruction (from a path) in such a way that an obstruction (a path) remains.

(20) W_k and W_l are, respectively, the α_kth and α_lth formulas of the decomposition; that is, they have the same meaning as in the list (4), but applied to (11).

Let us explain the content of these definitions (see also Lemma 13). If $Y_1, \cdots, Y_\sigma$ is an obstruction in Y, then one can specify a splitting process for Y (by means of reverse application of rules of **G** without preserving the principal formula in the rule $(\rightarrow \exists)$), by which one obtains exactly σ branches, the ith of which contains as component the formula arising from the occurring of Y_i in Y $(i = 1, \cdots, \sigma)$. If $Y_1, \cdots, Y_\sigma$ is a path, then in the splitting process for Y, one can obtain a branch whose top sequent contains exactly σ formulas, arising from $Y_1, \cdots, Y_\sigma$. For a formula of the propositional calculus, this definition is especially clear: such a formula is equivalent to the conjunction of all its paths and the disjunction of all its obstructions (see [20]).

A collection

$$[C, \beta_1; S_1], \quad [C, \beta_2; S_2], \ldots, [C, \beta_\nu; S_\nu] \tag{12}$$

of elementary extensions of sequence (6) is said to be *complete* if: 1) $S_1, \cdots, S_\nu$ are pairwise consistent; 2) for some l $(0 \leqslant l \leqslant h)$ the last component of sequence (12) is subordinate to the lth; and 3) the occurrences in (10) of the formulas $V_{\beta_1}, \cdots, V_{\beta_\nu}$ constitute an obstruction in V_{α_l} (in F for the case $l = 0$).

LEMMA 11. *Assume that* (12) *is a complete collection of extensions of sequence* (6). *Then every branch in which* (6) *occurs can be built upwards so that not more than* ν *branches are obtained, in each of which one of the sequences of* (12) *occurs.*

PROOF. Assume that (6) occurs in **B**, $l \neq 0$, and X_l is the lth component of the definition of a realization of the sequence in **B**. Let us consider an obstruction in X_l, the existence of which arises from the definition of a complete collection. This obstruction consists of certain occurrences of formulas $\overline{V}_{\beta_1}, \cdots, \overline{V}_{\beta_\nu}$ (the bar over the formulas signifies the appropriate substitution which affects variables which are simultaneously basic variables of the formula V_{α_l}). We observe the splitting process for X_l, turning our attention to the occurrences of $\overline{V}_{\beta_1}, \cdots, \overline{V}_{\beta_\nu}$. Let Y be the maximal formula occurring in X_l and containing any of these occurrences and such that the formula arising from Y remained as a component of the sequents until the top of the branch **B**. Y can coincide with X_l. If there is no such formula Y (that is, no suitable formula survived to the top—"it was split"), then this splitting process affected and "penetrated to the core" some V_{β_i}. Then, as is evident, there already occurs in our branch the sequence $[C, \beta_i; S_i]$, and no additional over-building is required.

Without limiting generality, we shall assume that Y contains $\overline{V}_{\beta_1}, \cdots, \overline{V}_{\beta_{\nu'}}$ $(\nu' \leqslant \nu)$. From the maximality of Y, Y does not lie within the scope of quantifiers of X_i (a formula beginning with $\exists$ certainly would be preserved upwards). Therefore Y occurs in the top sequent of the branch, and all basic variables of the formulas $V_{\beta_1}, \cdots, V_{\beta_{\nu'}}$ which are not basic variables of the formula V_{α_l} are not replaced by anything in this uppermost occurrence. On the other hand, basic variables in common with V_{α_l} are replaced as in X_l, which is required for subordinacy to the lth component of the sequence. Thus we can begin to split Y until the realization of the formulas $V_{\beta_1}, \cdots, V_{\beta_{\nu'}}$ is attained (in removing quantifiers it is necessary to choose values of the variables in a way similar to the way it was done in the proof of Lemma 3). As a result we obtain ν' branches, each of which contains a corresponding sequence of (12).

To complete the proof of the lemma, we note that the case $l = 0$ is analyzed similarly (but more simply); the role of Y is played here by one of the formulas $F_1, \cdots, F_n$, the one within which lies the obstruction being used. We remark that nowhere in the proof did we use the properness of the occurrences of $V_{\beta_1}, \cdots, V_{\beta_\nu}$ in V_{α_l}.

4. To complete the description of our concretization, we stipulate that Rules **A** and **Б** for constructing favorable sequences are formulated as in §1, but we begin with the following definition of a closed sequence: a sequence **H** is said to be *closed* if the corresponding formula is strongly equivalent to some closed disjunction.(21) This extension of the concept of closed sequence is not necessary for the completeness of the method (see Remarks 3 and 5).

Thus for sequents of arbitrary structure we now have the inductive concept of *favorable sequence.* The properties of these favorable sequences are completely similar to those presented in §§2.2 and 2.3. In fact, Lemmas 4–8 and the Corollary of Lemma 4 can be carried over to this case without changing their formulations and proofs (beforehand one needs a word-for-word extension of the notions of *foundation with respect to favorability*, *factorability*, *factor* of a sequence, the operation of *adjoining a sequence*, and *contraction*, *permutation*, and *specification* of a sequence). In particular, we obtain the correctness of the method: *from favorability of the empty F-sequence follows the derivability in* **G** *of sequent* (10). (Completeness of the method clearly requires some assumptions about the decomposition; for example, for the empty decomposition there are generally no favorable sequences.)

From the properties of favorable sequences we can draw the same conclusions about the possibility of limiting verifications as in §2. In particular, the modifications $\mathbf{A}^+$ and $\mathbf{Б}^+$ of Rules **A** and **Б** are convenient, as well as the transition to dc-foundations (with the same definition of *dc-applications* and *+-closed sequences*).

We note still another property of favorable sequences which is applicable in the general case more often than for sequents of the form (1). We shall say that **H** is a *null-fusion* of sequence (6) if it is obtained from (6) by excision with respect to the

(21) This means that some disjunction $D_1 \vee X \vee D_2 \vee \neg X \vee D_3$ can be transformed into the formula corresponding to H by "moving inwards" of negations. (For details about the concept of strong equivalence, see [2], §2.2.)

characteristic $1, \cdots, k-1, k+1, \cdots, h$, and for some l $(0 \leqslant l \leqslant h, l \neq k)$ the kth component of sequence (6) is subordinate to the lth and the occurrence of W_k in (10) forms an obstruction in the occurrence of W_l (the meaning of W_k and W_l is the same as in the list (4); $W_l = F$ for the case $l = 0$). If we move the kth component of sequence (6) backwards and apply Rule Б^+ with the resulting sequence as the unique argument, we arrive precisely at **H**. Thus we have

LEMMA 12. *A null-fusion of a favorable sequence is favorable.*

The lemma shows that, in the process of constructing favorable sequences, it is not necessary to obtain sequences admitting null-fusions (in particular, at the very beginning of the process, instead of +-closed sequences admitting null-fusions, one must begin immediately with an application of Rule Б^+ to these null-fusions). For sequents of form (1) there is possible only one form of null-fusion: $l = 0$, W_k coincides with some $D_{i,1}$, and $\delta_i = 1$. A corresponding special case of Lemma 12 is applied in the machine algorithm described in [13].

§6. Generalization of Rule Б. Completeness of the method

1. Let Γ be a collection of occurrences in the formula Y of some subformulas; we shall say that Γ *overlaps* a path Π if for each occurrence in Γ one can find an element of Π which is contained in this occurrence. We call a path *simple* if it consists of occurrences of elementary formulas. We shall say that a list

$$\Gamma_1, \ldots, \Gamma_\sigma, \tag{13}$$

consisting of collections of occurrences in Y, forms a *generalized obstruction* in Y if: 1) each simple path in Y is overlapped by some collection in (13), and 2) condition 1) is violated if we eliminate from (13) any of its components.(22) For example, each of the lists $\{A, B\}$, $\{C, D\}$, $\{F, G\}$, and $\{E\}$, $\{B, G\}$, $\{C, F\}$ forms a generalized obstruction in the formula

$$\frac{A \vee B}{C} \vee \frac{D \vee E}{F} \vee G.$$

The collection $\{B, C\}$ does not overlap any path in this formula, and therefore it cannot enter into any generalized obstruction.

LEMMA 13. *The definition of a generalized obstruction for the case where all collections of* (13) *are one-termed is equivalent to the definition of obstruction.*

(22) For practical detection of generalized obstructions it is necessary to check not just simple paths, but rather "large-scale" ones (but, nevertheless, not containing proper superformulas of formulas occurring in (13)). Overlapping of such a path implies overlapping of the whole group of corresponding simple paths.

PROOF. Let $Y_1, \cdots, Y_\sigma$ be an obstruction in Y. We shall prove by induction with respect to the logical length of Y that any simple path is overlapped by some of the elements of the obstruction. If Y is elementary, the assertion is obvious. Just as obvious is the case $Y = Y_1$ (and $\sigma = 1$). Assume now that $Y \neq Y_1$. Removal of a quantifier does not change the situation. Assume $Y = A \mathbin{\&} B$. Then any simple path is related either to A or to B, and the elements of the obstruction going to A and B, respectively, form obstructions in their formulas. Consequently, by the inductive hypothesis, both the paths from A and the paths from B are overlapped. Assume $Y = A \vee B$. Then the given obstruction entirely leads out, for example, to A, and the elements going out to A in any simple path form a path in A. The inductive hypothesis is again applicable, completing the proof that $Y_1, \cdots, Y_\sigma$ satisfies the first condition for generalized obstructions. The second condition also is fulfilled: eliminating Y_1, for instance, from the obstruction, we easily construct a simple path (passing through Y_1), not overlapped by the remaining elements $Y_2, \cdots, Y_\sigma$.

By a similar induction one proves the inverse implication. We shall mention only the least trivial step. Assume $Y = A \vee B$. Can a generalized obstruction (13) with single-element Γ_i's go out partially to A and partially to B? Let us suppose that all simple paths in B are overlapped. Then they also are overlapped in Y, and, moreover, without the help of elements of A. Hence, by the second condition on generalized obstructions, A does not contain elements of (13). If some path in B is not overlapped, then all simple paths in A are overlapped (otherwise one could arrange a simple path, not overlapped in Y). Then, similarly, B does not contain elements of (13). Thus we find ourselves with the conditions of applicability of the appropriate inductive hypothesis, and we can complete the proof.

LEMMA 14. *Eliminating from the components of the obstruction* (13) *any number of occurrences which are part of* $\Gamma_1, \cdots, \Gamma_\sigma$, *we obtain a list* $\Gamma'_1, \cdots, \Gamma'_\sigma$ *for which at most the second condition on generalized obstructions is violated.*

One can keep in mind that $\Gamma'_1, \cdots, \Gamma'_\sigma$ do not degenerate into the empty collection, nor does the empty collection overlap any path. The proof of the lemma is obvious.

2. The collection

$$[C, \beta_1^1, \beta_1^2, \ldots, \beta_1^{h_1}; S_1], \quad [C, \beta_2^1, \beta_2^2, \ldots, \beta_2^{h_2}; S_2], \ldots, [C, \beta_\nu^1, \beta_\nu^2, \ldots, \beta_\nu^{h_\nu}; S_\nu] \quad (14)$$

of extensions of sequence (6) is said to be *complete* if it satisfies the following conditions:

1) $S_1, \cdots, S_\nu$ are pairwise consistent.

2) For some l $(0 \leqslant l \leqslant h)$ and for every i and j $(1 \leqslant i \leqslant \nu; h < j \leqslant h + h_i)$ the jth component of the sequence $[C, \beta_i^1, \cdots, \beta_i^{h_i}; S_i]$ is subordinate to the lth component of this sequence.

3) The collection of occurrences in F of formulas $V_{\beta_i'}, \cdots, V_{\beta_i^{h_i}}$ $(i = 1, \cdots, \nu)$ is the ith of the collections of some ν-termed generalized obstruction in V_{α_l} (in F for the case $l = 0$).

It is useful to take into account that, for verification of overlapping of paths in V_{α_l}, it suffices to consider only paths composed of proper occurrences in V_{α_l} of formulas of the decomposition (see footnote 22).

RULE $\mathbf{Б}_\Pi$. Assume that $\mathbf{H}_1, \cdots, \mathbf{H}_\nu$ are favorable sequences and sequence (6) possesses the following property: one can find a complete collection (14) of extensions of (6) such that, for all i $(1 \leqslant i \leqslant \nu)$, $\mathbf{H}_i$ is a subsequence with respect to a characteristic of the form $j_1, j_2, \cdots, j_{h'}, h+1, h+2, \cdots, h+h_i$ (here $h' \leqslant h$ and $1 \leqslant j_1 < \cdots < j_{h'} \leqslant h$) of the ith sequence of (14).(23) Then (6) is considered a favorable sequence.

LEMMA 15. *Every application of Rule* $\mathbf{Б}_\Pi$ *can be replaced by a sequence of applications of Rule* $\mathbf{Б}$ *leading to the same sequence.*

It suffices to prove the lemma for the case where, for all i $(1 \leqslant i \leqslant \nu)$, $\mathbf{H}_i$ coincides with the ith sequence in the list (14). We proceed by induction with respect to the quantity $h_1 \cdot h_2 \cdot \cdots \cdot h_\nu$. If $\Pi_1^\nu h_i = 1$, then the given application is an application of Rule $\mathbf{Б}$ (by virtue of Lemma 13). Assume $h_1 > 1$. We replace the given application of $\mathbf{Б}_\Pi$ by the following two:

1) The arguments of the first are the sequences of (14), and the result is the sequence $\mathbf{H}$ obtained from $[C, \beta_1^1, \cdots, \beta_1^{h_1}; S_1]$ by excision with respect to the characteristic

$$1, 2, \ldots, h, h+1, \ldots, h+h_1-1.$$(24)

2) The arguments of the second are $\mathbf{H}$ and those sequences of (14) for which $\mathbf{H}$ is not a subsequence, and the result is sequence (6).

Both of the applications described are possible: the consistency of the corresponding bundles follows from the pairwise consistency of $S_1, \cdots, S_\nu$ (no new relations not occurring in $S_1, \cdots, S_\nu$ arise), the required subordinacy to the lth component is not violated, and a violation of the generality of the obstructions can, by virtue of Lemma 14, be corrected by throwing out unnecessary arguments (see footnote 24).

(23) The last h_i components of the sequence $\mathbf{H}_i$ and of the ith sequence of (14) are called the *outer components* of the application of Rule $\mathbf{Б}_\Pi$.

(24) We have in mind that not every one of the sequences of (14) need be an argument of the described application. In fact, if Π is a path overlapped by an occurrence of formula $V_{\beta_1^{h_1}}$, but not overlapped by the whole collection of occurrences $V_{\beta_1^1}, \cdots, V_{\beta_1^{h_1}}$ (in the general case, the set of such paths is nonempty), then those sequences of (14) which overlap Π turn out to be superfluous. This footnote is also related to the second application.

By the inductive hypothesis, each of them can be replaced by a sequence of applications of Rule **Б** (the inductive parameter of the second application has decreased, since $\mathbf{H}_1$ is not among its arguments). The lemma has been proved. We remark that the total number of required applications of Rule **Б** does not exceed $\Pi_1^\nu h_i$ (moreover, in the absence of repetitions among the numbers β_j^i this bound is realized).

Thus along with Rule **Б** we have obtained the possibility of using for the construction of favorable sequences the more "large-scale" Rule $\mathbf{Б}_\Pi$. In spite of the fact that, using $\mathbf{Б}_\Pi$, we can arrive at the empty sequence, generally speaking, in a smaller number of steps, one can hardly recommend the constant use of this more laborious rule. In all cases where it is possible to use intermediate results obtained by a sequence of applications of Rule **Б**, which replaces some application of $\mathbf{Б}_\Pi$, it is not advisable to use $\mathbf{Б}_\Pi$. However, in certain cases, one can guarantee that a realization of a sequence-argument of an application of $\mathbf{Б}_\Pi$ in a branch possesses the following property: realizations of the outer components of the application arise from a direct splitting of the formula of the realization of the corresponding component of this sequence (which turns into the lth component of the resulting sequence; for $l = 0$ we are talking about a direct splitting of some one of the formulas $F_1, \cdots, F_n$), not requiring repeated removal of the same quantifiers. Verification of these cases is connected with regular* derivations (see [1], §7.2) and the prohibition of sequences which can occur only in branches of irregular trees. These "irregular" sequences are easily recognized, and, upon their appearance (which, generally speaking, is unavoidable when using only **Б**), it is reasonable to apply $\mathbf{Б}_\Pi$, which jumps immediately over all these "irregular" sequences. This is even more advisable, as the laboriousness of the application of $\mathbf{Б}_\Pi$ is, in the cases described, insignificantly different from that of a single application of **Б** to the same sequence-arguments. The occurrence of these cases is especially characteristic of sequents obtained by elimination of positive quantifiers.

For the proof of Lemma 15 we used Rule **Б**; if we limit ourselves to applications of $\mathbf{Б}^+$, we can guarantee only that one will obtain some subsequence of the sequence resulting from the given application of $\mathbf{Б}_\Pi$. The equality of this subsequence to the result cannot always be ensured, even if we consider, instead of $\mathbf{Б}_\Pi$, its natural modification $\mathbf{Б}_\Pi^+$ (obtained by adding the requirement of nonexistence of proper subsequences possessing the same property as the result).

REMARK 4. In terms of sequential calculi, the result contained in Lemma 15 is analogous to the following assertion: on sequents of form (10), not containing $\vee$ as outer symbol, the calculus **G** is coextensive with the calculus obtained by replacing the rule $(\to \vee)$ by the rules

$$\frac{\to \Gamma,\ A,\ \Gamma'}{\to \Gamma,\ A \vee B,\ \Gamma'},\qquad \frac{\to \Gamma,\ B,\ \Gamma'}{\to \Gamma,\ A \vee B,\ \Gamma'}.$$

---

*Editor's note. The Russian term is правильный; it was translated as "correct" in the English version of [1].

3. Before considering the problem of the completeness of the method, we introduce certain normalizations of proofs in **G**. A proof in **G** of sequent (10) is said to be *weakly minus-normal* if in any application of $(\rightarrow \exists)$ the terms $t_1, \cdots, t_m$ are F-terms (compare the requirement of minus-normality in, for example, [21], [22]). We shall call a proof in **G** *semicomplete* if it satisfies the following restriction: if in the sequent-conclusion of some application of rule $(\rightarrow \exists)$ there is a formula with principal symbol & or $\vee$, then nowhere in the tree above was this formula used as principal formula; that is, under the transition to a "weeded-out" proof(25) this formula would be weeded out, since neither it nor results of its "splitting" is used in the tree to obtain axioms. It is easily proved that the requirements of semicompleteness and weak minus-normality do not change the class of sequents derivable in **G**.

As we shall see below, the complexity of a foundation of favorability of the empty F-sequence is determined by the complexity of a derivation of sequent (10). In the proof of the completeness of the method we shall assume that the normalizations mentioned above are satisfied. Therefore it is important that this assumption not lead to a significant complication of the derivations in **G**. Weak minus-normality (in contradistinction to minus-normality) always can be introduced into a proof in such a way that it only becomes simpler. Somewhat worse in this respect is the requirement of semicompleteness: in transforming an arbitrary derivation $\mathfrak{D}$ into a semicomplete derivation $\mathfrak{D}'$ the total number of sequents in the derivation can increase. However, if $\mathfrak{D}$ did not contain "unnecessary" applications of rules (that is, applications which would be weeded out), then the complexity of $\mathfrak{D}'$ is equivalent in a certain sense to the complexity of $\mathfrak{D}$. This assertion can be made more precise in terms of the number of paths that have to be examined in the tautological Herbrand development of the original sequent. In fact, the weeded-out tautological developments constructed from $\mathfrak{D}$ and $\mathfrak{D}'$ coincide (up to permutations of components of disjunctions and conjuntions).

4. Let us consider a requirement on a decomposition, guaranteeing the completeness of the method. Assume X is some occurrence in the formula Y. We shall say that Y is a *disjunctive* (*purely disjunctive*) *superformula* of this occurrence if X does not lie within the scope of any symbol & (respectively, symbol & and quantifiers) occurring in Y. A decomposition (11) of sequent (10) is said to be *complete* if for every occurrence X in (10) of an elementary formula one can find a formula of the decomposition whose occurrence in (10) is a purely disjunctive superformula of X. As we shall see below, for any complete decomposition and derivable sequent (10), the empty F-sequence is favorable; but the complexity of a foundation of its favorability can turn out to be the same as it is under a preliminary reduction of all formulas

(25) For the concept of a *weeding-out*, see [23], [24]; a weeded-out proof does not contain superfluous applications of rules (compare Lemma 3 of [11]). The concept of a *weeded-out Herbrand decomposition*, used below, has a similar meaning.

$F_1, \cdots, F_n$ to prenex form. Hence the introduction into the decomposition of "large-scale" formulas (although not needed for the completeness of the method) permits the advantageous use of nonprenex sequents in establishing deducibility.

We call an occurrence of a formula $\exists xX$ in F an $\exists$-occurrence if it is not an occurrence in a formula of the form $\exists x'\exists xX$ and $\exists xX$ does not coincide with F_i $(i = 1, \cdots, n)$. The result of removing from a decomposition certain formulas will be called its *constriction* (we have in mind that, for the remaining formulas, their occurrences in (10) are determined in the same way as before the constriction). We note that, upon constriction of a decomposition, some possibilities for applying Rules **A**, **Б** and $\mathbf{Б}_\Pi$ disappear, but some new ones can appear (since improper occurrences can become proper).

LEMMA 16. *Let* $\mathfrak{D}$ *be a semicomplete derivation of sequent* (10), *and let* (11) *be a complete decomposition of sequent* (10) *containing every formula whose occurrence in* F *is an* $\exists$*-occurrence. Then one can construct a constriction of the decomposition in such a way that there is a foundation of favorability of the empty* F*-sequence containing no more applications of* $\mathbf{Б}_\Pi$ *than there are applications of the rule* $(\rightarrow \exists)$ *in* $\mathfrak{D}$.

PROOF. Assume that $\mathfrak{D}$ is weeded out and weakly minus-normal, and that all axioms in $\mathfrak{D}$ have the form

$$\rightarrow \Gamma, \ni, \Gamma', \ni', \Gamma'' \qquad (15)$$

where $\ni$ and $\ni'$ are elementary formulas, of which one is the negation of the other. If a preliminary transformation into a derivation satisfying these requirements is needed, then it is achieved by means of a weeding out, the replacement of certain terms, and the addition above of certain applications of the rule $(\rightarrow \vee)$. Moreover, the number of applications of the rule $(\rightarrow \exists)$ can only decrease. We eliminate from the decomposition every formula X for which there is a formula of Y whose occurrence in (15) is a purely disjunctive superformula of the occurrence of X; we shall say that the decomposition has no *superfluity*.

The proof is divided up as in §2.1. In a similar way, for every branch **B** one determines an *equivalent* sequence **H(B)** which itself occurs in **B** but has no proper supersequences occurring in **B**. Every branch **B** of the tree $\mathfrak{D}$ is equivalent to a closed sequence. In fact, assume the uppermost sequent of the branch is (10). In view of the completeness of the decomposition, before obtaining both $\ni$ and $\ni'$ we must generate some purely disjunctive superformulas $\overline{W}_1$ and $\overline{W}_2$ of these formulas, whose preimages W_1 and W_2 (see p. 41) entered into the decomposition ($\overline{W}_1$ can coincide with $\overline{W}_2$); hence **H(B)** contains components corresponding to these formulas. Then the formula corresponding to **H(B)** contains $\overline{W}_1$ and $\overline{W}_2$ and is a closed disjunction.

Let $\mathfrak{D}'$ be the result of cutting off $\mathfrak{D}$ above certain sequents and assume that the *basic property* is satisfied: every branch of $\mathfrak{D}'$ is equivalent to some favorable sequence. Let us consider a branch **B** of $\mathfrak{D}'$, one of those in which the number of applications of the rule $(\to \exists)$ is maximal. Assume Σ is the uppermost sequent in **B**. We descend from Σ to the first application of $(\to \exists)$. Because of the semicompleteness and the weeding out, there is a side-formula W in the premise Σ' of this application (Σ' can coincide with Σ) such that in the tree above Σ' there occur only splittings of W according to the rules $(\to \&)$ and $(\to \vee)$ (from the choice of **B** there are also no applications of $(\to \exists)$ above Σ' in all other branches passing through Σ').[26] The following cases are possible:

1) Between Σ' and Σ in **B** there stands Σ_0, in which some W_0 occurs, so that only splittings according to the rule $(\to \vee)$ occur between Σ_0 and Σ, and the preimage of W_0 is a formula of the decomposition.

2) This coincides with clause 1) except that W_0 splits according to the rules $(\to \vee)$ and $(\to \&)$.

3) There is no such sequent Σ_0.

In the first case, above Σ_0 in **B** there does not arise any realization of components of **H(B)** (because the decomposition has no superfluity); therefore everything above Σ_0 can be cut off without violating the basic property of the tree $\mathfrak{D}'$. Let us consider the third case. Since all $\exists$-occurrences are mentioned in the decomposition, there is below Σ' in **B** an occurrence of formula W_0 in some sequent, which is a realization of some component (say, the lth) of the sequence **H(B)** and such that the reverse application to W_0 of the rule $(\to \exists)$ yields W. A special subcase arises when this W_0 turns out to be one of the formulas $F_1, \cdots, F_n$, i.e. it is not a realization of any component of the sequence **H(B)**. In this subcase we assign 0 to be the number l mentioned above. We denote by Σ_0 the sequent lying under Σ' (W_0 occurs in Σ_0).

In the second, as well as in the third, case we cut out of $\mathfrak{D}'$ everything standing above Σ_0. We denote by **H** the sequence equivalent to the branch leading to Σ_0. The sequences equivalent to the deleted branches form a complete collection of (generally speaking, nonelementary) extensions of **H**. In this connection, the l from the definition of a complete collection coincides in the third case with the l mentioned above. For the second case, l is chosen similarly ($l \neq 0$). The overlapping of every path of the definition of a generalized obstruction arises from the fact that all the necessary results of the splitting of W (W_0 in the second case) went out in the deleted branches. The required pairwise consistency of the bundles also is not disturbed, since all relations for all atoms were known earlier for $\mathfrak{D}$ and are consistent. Thus **H**

---

(26) Without disturbing the basic property of $\mathfrak{D}'$ and the number of applications of $(\to \exists)$ in it, one can choose any sequence of splittings of W.

can be obtained from one application of Rule Б_Π; hence we again have a tree for which the basic property holds.

Repeating the process just described, we finally reach a tree composed of a single sequent, the sequent (10); that is, we have verified the favorability of the empty sequence. Let us now consider the problem of the number of applications of Б_Π which have been used. Constricting the decomposition to simple disjunctions and formulas of the form $\exists xX$, we can eliminate the second of the indicated cases, and, in the third case, application of Б_Π in an induction step just corresponds to the elimination from $\mathfrak{D}$ of one application of $(\to \exists)$. This completes the proof of the lemma.

THEOREM 5. *Let* (11) *be a complete decomposition such that, for every* $\exists$-*occurrence in* F, *one of its disjunctive superformulas lies in* (11) *or coincides with* F. *For every semicomplete derivation* $\mathfrak{D}$ *of sequent* (10) *one can construct a foundation of favorability of the empty* F-*sequence. If, in addition,* (11) *consists of simple disjunctions and formulas of the form* $\exists xX$, *then this foundation contains no more applications of* Б_Π *than there are applications of* $(\to \exists)$ *in* $\mathfrak{D}$.

The proof coincides with the proof of Lemma 16 except in the definition of formula W_0 in the third case of the cases described in the lemma. One must choose a W_0 whose splitting by the rules $(\to \vee)$ and $(\to \exists)$ leads to W (moreover, in the branch **B** these splittings can be alternated with applications of rules with other principal formulas, not coming out of W_0). The requirements on the decomposition assumed in the theorem guarantee the existence of such a formula W_0 (though it may not guarantee, as in the lemma, that W_0 is separated from W by one application of $(\to \exists)$). For the consideration of the first case in the lemma, we base ourselves upon the nonsuperfluity of the decomposition. We cannot immediately use a constriction; therefore we note that in the proof of the first half of the theorem we can consider the first case as a special case of the second (and not go over to a decomposition without superfluity). However, for the proof of the second half of the theorem we must get rid of the second case of the lemma. On the other hand, the condition on existence in the decomposition of only simple disjunctions and formulas $\exists xX$ is such that the nonsuperfluity can turn out to be violated only on simple disjunctions. Having in the decomposition a simple disjunction D and one of its subformulas D', we always can use only D in sequences; that is, we can construct a foundation as if D' were not present in the decomposition, which is equivalent to a transition to a constriction, but does not require its formal execution.

The theorem is proved. We note that the use of disjunctive superformulas of $\exists$-occurrences permits not only a decrease in the total number of components of the decomposition but also permits us to replace in certain cases a whole chain of applications of $(\to \exists)$ by one application of Б_Π. For example, the two applications of $(\to \exists)$ to remove the quantifiers in $\exists x\ (\exists x'X \vee Y)$ require one Б_Π. It is true that whether the following figure in $\mathfrak{D}$

$$\begin{array}{ll} \to \Gamma,\ [X]^{x,\, x'}_{t,\, t'}\ [\exists x' X]^{x}_{t},\ [Y]^{x}_{t},\ \exists x\, (\exists x' X \vee Y),\ \Gamma' & \\ \hline & (\to \exists) \\ \to \Gamma,\ [\exists x' X]^{x}_{t},\ [Y]^{x}_{t},\ \exists x\, (\exists x' X \vee Y),\ \Gamma' & \\ \hline & (\to \vee) \\ \to \Gamma,\ [\exists x' X \vee Y]^{x}_{t},\ \exists x\, (\exists x' X \vee Y),\ \Gamma' & \\ \hline & (\to \exists) \\ \to \Gamma,\ \exists x\, (\exists x' X \vee Y),\ \Gamma' & \end{array} \qquad (16)$$

reduces to one or two applications of $\mathbf{Б}_{\Pi}$ depends upon whether or not one uses $[Y]^{x}_{t}$ in the tree above.

COROLLARY. *Completeness of the method holds for complete decompositions.*

In fact, a sequent $\to F'_1, \cdots, F'_n$ obtained from (10) by taking out the quantifiers is equivalent to (10). For any complete decomposition of (10), the corresponding decomposition of the sequent $\to F'_1, \cdots, F'_n$ satisfies the requirements of the first half of the theorem.

REMARK 5. In the proofs of this section, we used only those closed sequences for which the corresponding disjunction is closed (but not strongly equivalent to a closed sequence). Thus the definition of closed sequence that we have assumed is not necessary for the completeness of the method, but can be used to increase the effectiveness of the corresponding algorithm (see Remark 3). The increase in effectiveness turns out to be especially noticeable for establishing deducibility in theories: in theorems and axioms of a theory one will encouter comparatively often various occurrences of the same formula (expressing some concept which is important in the theory), on account of which there will arise complicated closed sequences (involving positive and negative occurrences of such formula-concepts).

§7. Extension of tactics

1. Let us consider an application of Rule $\mathbf{Б}^+$, in the notation of p. 18. We shall say that the kth component of sequence (6) is a *descendant* of the kth component of the sequence $\mathbf{H}_i$ $(1 \leqslant i \leqslant \nu)$ if, in the characteristic $j^i_1, \cdots, j^i_{h_i}$ with respect to which $\mathbf{H}_i$ is a subsequence of $\mathbf{H}'_i$, $j^i_{k'} = k$.(27) In the formulation of the method of §1, a distinctive property was that every component of a sequence resulting from an application of $\mathbf{Б}^+$ is a descendant of something or other. In the general case one can find components of sequence (6) which are not descendants of anything (such a component might turn out to be the lth component of this sequence, where l is taken from the definition of a complete collection of elementary extensions and $l \neq 0$; see p. 42). In the latter case we shall say that the corresponding component of the

(27) This definition is formulated under the assumption that the resulting sequence has not undergone a trivial contraction. For every component of the result of a trivial contraction, it is, in turn, easy to show whose descendant it is. If the descendant of the k'th component of the sequence $\mathbf{H}_i$ is eliminated under a trivial contraction, then we shall say that this component has a *degenerate descendant.*

resulting sequence is a descendant of the outer components of the application of Б^+. The existence of descendants of outer components does not introduce any change in the definitions of the dis-, ss-, П- and sup M-tactics.

Analysis of the results of §3 shows that all the proofs carry over without change to the general case if we admit only complete decompositions. We note that a new factor in the proof of Lemma 9 is the following: in tracing an element ξ through an improper part, it is not always possible to ascend to sequences which are boundaries of proper and improper parts. However, the occurrence of this case in no way affects the required transformations. The choice in this proof of the sequences $\mathbf{H}^{\xi}$ as the lowest sequences guarantees, even in the general formulation of the method, the impossibility of occurrences of the element ξ below occurrences of these sequences. Thus we have

THEOREM 2′. *For any derivable sequent* (10), *for any complete decomposition of this sequent, and for any acyclic ordering* П, *one can construct a dc-foundation of favorability of the empty sequence which satisfies the* П-, *dis- and ss-tactics.*

Let us note that, as the hypothesis for this result, we really require not the completeness of the decomposition, but rather the existence of some foundation of favorability of the empty sequence.

2. The situation is somewhat more complicated in the case of the support tactic. Let us consider the sequent

$$\rightarrow \neg Pa,\ \exists x\,(Px\&F'),$$

where $[F']^x_a$ is some derivable formula not containing the predicate P. Assume that a complete decomposition of this sequent contains the formulas $\neg Pa$, Px and F'. We choose as support set M the single formula $\neg Pa$. It is easily seen that such a support set is necessary, but to obtain the empty sequence it will be necessary to make at least one application of Б, using only closed F'-sequences. Hence a direct carry-over to the general case of Theorem 3 (and consequently also of Theorem 4) is impossible.

THEOREM 3′. *For any derivable sequent* (10) *and any complete decomposition of this sequent, one can find a constriction of the decomposition such that, for any necessary* M (M *is found by constriction*), *the combination of the* sup M-, ss- *and dis-tactics is complete.*

In fact, if one takes as constriction a decomposition containing only simple disjunctions (including single-termed ones), then the proof of §4 is completely preserved. For such a constriction, descendants of outer components are impossible; therefore the verbatim definition of the eng-tactic (in the general case it is natural to permit, for results of eng-applications, the existence of such descendants along with descendants of the principal argument and descendants of closed sequences) leads us to the appropriate analogue of Theorem 4. We note that the applied constriction is, of course, too stringent.

Analogues of Theorems 3 and 4 are possible which do not require complete elimination of "large-scale" formulas of the decomposition. More suitable forms of analogues of these theorems from a practical viewpoint (and generally not requiring transition to constrictions) are obtained in terns of the *compound rule* **Б** which is defined below.

A tree $\mathfrak{D}$ of applications of Rule $\mathbf{Б}^+$ is said to be an *application of a compound rule*, the *arguments* of which are considered to be the sequences which do not lie under any application of $\mathbf{Б}^+$, if the outer components of any application in $\mathfrak{D}$ either occur in an argument of the compound rule or are descendants of outer components of applications which lie above. The lowest sequence of the tree $\mathfrak{D}$ is called the *result* of the application of the compound rule, and the last components of the arguments of the compound rule are called the *outer components* of the application of this rule.

LEMMA 17. *Assume* **R** *is a constriction of a decomposition* (11) *of sequent* (10). *Then every application of Rule* $\mathbf{Б}^+$ *for sequences constructed on the basis of the decomposition* **R** *can be replaced by an application of the compound rule* (*in a formulation based upon the decomposition* (11)) *with the same arguments, outer components, and result.*

The proof is obvious for the case where the resulting sequence of the original application of $\mathbf{Б}^+$ is not subjected to a trivial contraction. In fact, if we look at this application from the point of view of the decomposition (11), then at most the condition of properness of occurrences is violated in it. To restore this properness it suffices to construct some intermediate applications of $\mathbf{Б}^+$, in each of which the descendant of the outer components must stand in the last place (and consequently is cut off in the following application). For the use in the original application of a trivial contraction, we construct first a suitable application of the compound rule for arguments from which are eliminated components having degenerate descendants in the original application. Then, introducing the eliminated components, we can observe that in the course of the application of the compound rule all of them are "inserted" somewhere (but not necessarily in those components in which they were inserted in the original application).

For the proof of the lemma, the only important thing is the possibility of placing the descendant of the outer components at the end of the sequence. Therefore, if the formulation of the tactic does not assume any restriction on the order of the components of the sequences and if the completeness of the tactic has been proved for constrictions of decompositions, then the obvious extension of this tactic to arbitrary decompositions (under which the role of Rule $\mathbf{Б}^+$ of the original tactic must be played by the compound rule of the generalization) also will be complete. This permits us to formulate convenient analogues of the results of §4 (but not the result about the combinability of the sup M-tactic with a tactic which introduces an ordering of components of sequences; see Theorem 6 below).

§8. Decompositions without repetitions. Weak Π-tactic

1. In §5.3 we agreed to consider decompositions with repetitions of formulas in order to be able to establish, for a formula of the decomposition, its occurrence in the original sequent. This convention lengthens the decompositions under consideration, and, more important, forces us to generate and use separately sequences which are obtainable from each other only by replacing indices of formulas of the decomposition by indices of equal formulas. Fortunately this convention is suitable only for the introduction of definitions, but has no effect on the essence of the method and its results. In fact, the content of §§5–7 can be preserved under a formulation of the method according to which, for a formula of the decomposition, there is determined a finite number of its possible occurrences in the original sequent. The concept of sequence in the new formulation can be considered as a "schema" of the former sequences (the concept of "sequence schema"** has the same meaning as "axiom schema"; for more details, see [1], §5, Remark 7). For example, if in the new decomposition the αth formula is equal to the α_1th, $\cdots$, α_sth formulas of the old decomposition, then the new sequence $[\alpha; S]$ is a schema, special cases of which are each of the former sequences $[\alpha_i; S]$ $(i = 1, \cdots, s)$.

We shall dwell only on certain details which are characteristic of the new formulation of the method. One and the same formula of the decomposition can now have both proper and improper occurrences in the comprehensive formula of the decomposition. Therefore, wherever we use this properness, it is necessary to fix the occurrence about which we are speaking. In the definition of *properness* it is reasonable to add to W the following condition: for properness of an occurrence X (of the formula W) in a formula W' it is necessary that one cannot find a disjunctive superformula of any occurrence Y of the same formula W which would contain X but which is not a disjunctive superformula of X.

EXAMPLE. Assume that the formula of the decomposition has the form

$$F_1 \vee \exists x (((\exists y M \vee F_2) \,\&\, (F_1 \vee F_3)) \vee (F_2 \,\&\, F_3)) \tag{17}$$

and the decomposition includes, of the explicitly written formulas, only F_1, F_2, F_3 and $(\exists yM \vee F_2)$ (from the convention on the form of sequent (10) it follows that F_1, F_2 and F_3 are quantifier-free). The proper occurrences of the indicated formulas of the decomposition in (17) are only the left occurrence of F_1, the right occurrence of F_2, both occurrences of F_3, and the occurrence of $(\exists yM \vee F_2)$. The left occurrence of F_2 does not satisfy the previous condition of properness (there is an "intermediate" formula of the decomposition), and the right occurrence of F_1 violates the

*******Editor's note.* The Russian expression is схема наборов; it was translated as "collection schema" in the English version of [1].

condition introduced just now ((17) is a disjunctive superformula of the left occurrence of F_1 containing a disjunctive superformula $F_1 \vee F_3$ of the second occurrence of F_1).

It is easily seen that any improper occurrence which violates the condition on properness which we have just introduced can be eliminated by use of an equivalence of the type $X \vee Y \equiv X \vee Y'$, where Y' is obtained from Y by replacing some occurrence of a formula of the form $(X \& B) \vee A$ by the formula A (in our example, $F_1 \vee F_3$ can be replaced by F_3, and the resulting formula will be equivalent to the original one). As a result of a series of such transformations, the formula F can be transformed into an equivalent formula F' such that the new condition is always fulfilled (for any selection of decompositions and any choice of the pair W and W'). It is essential in this connection that a proof of F' is never more complicated than a proof of F. Therefore, instead of introducing the indicated condition of properness, one can assume that a suitable preliminary standardization has been carried out.

Let us note another special property of the new formulation. Earlier, in considering a complete collection of extensions (12), all of the numbers $\beta_1, \cdots, \beta_\nu$ turned out to be different. Now some of these may coincide (and consequently repetitions are possible in the list (12)). If, for example, $\beta_1 = \beta_\nu$, then, in constructing obstructions in V_{α_l} (in F for the case $l = 0$), it is necessary to select two different proper occurrences of $\overline{V}_{\beta_1}$ and $\overline{V}_{\beta_\nu}$ (of the formula V_{β_1}) in V_{α_l} (among other things, under these conditions V_{β_1} must be quantifier-free). As a result, in the formulation of **Б** (and $\mathbf{Б}_\Pi$) there arises the necessity of permitting repetitions of arguments (which is perfectly natural in terms of "sequence schemas").

2. As Theorem 5 shows (also see what was said about figure (16)), for economization of the decomposition and shortening of the process of establishing deducibility it is advisable to include in the decomposition a maximal disjunctive superformula of the elementary formulas. This holds for the previous formulation of the method and, to a certain extent, remains true for decompositions without repetitions. However, now, in connection with the possibility of simultaneous use of identical formulas occurring at different places in the sequent, it turns out to be convenient to have frequent repetitions of occurrences in the sequent of one and the same formula of the decomposition. This in no way contradicts what has been said above with respect to superformulas of $\exists$-occurrences, but forces us to reexamine the relationship of elementary formulas to superformulas. For every given sequent, the most convenient choice of a decomposition presents a special problem (see [1], §3.3), but if we assume a uniform way of constructing complete decompositions of sequent (10), then apparently it is convenient to agree on the following:

In a decomposition of sequent (10) once includes all elementary formulas which occur positively in (10) and all maximal disjunctive superformulas of $\exists$-occurrences.(28)

REMARK 6. Under such a choice of decompositions, we count on the use of the same elementary subformulas of the original sequent. For greater convenience in such a use one could propose another language for presenting sequences (see Remark 7 of [1]). Such a language for sequences, consisting of elementary formulas, is close to a language of the method of resolution (see [6], §3).

3. For the remainder of this section, by decompositions we shall understand complete decompositions consisting of elementary formulas (carrying over the following result to arbitrary decompositions would be of little convenience; see the end of §7). The concepts of the inverse method that we shall use will be assumed to be formulated on the basis of such decompositions.

We shall say that a foundation of favorability of some sequence satisfies the *weak* Π-*tactic* (*w*Π-*tactic*) if all sequences in this foundation, except the closed ones, are Π-ordered. The concept of wΠ-tactic is, among other things, of interest in connection with a corresponding tactic of the method of resolution (more precisely, the clash method, for which the most natural analogue of the tactic, unifying the ordering of components of a sequence, is just the wΠ-tactic; see §6 of [6]). Moreover, many tactics are essentially based upon the possibility of permuting components only in closed sequences; hence the wΠ-tactic remains complete in combination with such tactics.

THEOREM 6. *For any derivable sequent* (10), *for any necessary support set* M *and any acyclic ordering* Π, *one can construct a dc-foundation of favorability of the empty sequence, satisfying the* wΠ-, sup M- *and ss-tactics.*

The proof is based upon the following modification of Lemma 10 (the meaning of the notation used below is the same as in §4.1).

LEMMA 18. *Assume* $\mathbf{H}$ *is not* $\{M^c\}$*-derivable, but the sequence* $\mathbf{H}\cdot\mathbf{H}^1$ *is* $\{M^c\}$*-derivable. One can construct a tree of applications of* $\mathbf{Б}^+$ *and* $\mathbf{Sp}$ *which is a foundation of favorability of* $\mathbf{H}$ *with respect to sequences in* M_{ix} *and in which the following conditions are satisfied*:

1) *In generating each nonclosed sequence, sequences of* M_{ix} *are involved.*
2) *In applications of* $\mathbf{Б}$, *supersequences of closed sequences do not arise.*
3) $\mathbf{H}^1$ *enters into the foundation only as part of closed sequences.*

The proof is constructed as in Lemma 10; moreover, we never resort to an

---

(28) Just such an algorithm for constructing decompositions is essentially assumed in §6 of [12] and §6.4 of [1]; in the terminology of those papers, nonelementary formulas of a decomposition are called *F-zones.*

extension of $\{\overline{M}\}$; that is, the role of $\{\overline{M}\}$ is always played by $\{M^c\}$. In this connection one makes use of the following fact (holding only for decompositions of elementary formulas): if $\mathbf{H}^1 \cdot \mathbf{H}^\beta$ is closed, then $\mathbf{H} \cdot \mathbf{H}^\beta$ is not $\{M^c\}$-derivable (otherwise $\mathbf{H}$ would also turn out to be $\{M^c\}$-derivable). Let us consider Case 2 of Lemma 10 (in Case 1 the proof remains unaltered). If, in addition, $\mathbf{H} \cdot \mathbf{H}^1$ is closed, then the third of the properties required in Lemma 18 is automatically fulfilled. Therefore to construct the required foundation is even simpler than in the case where $\mathbf{H} \cdot \mathbf{H}^1$ is not closed.

Assume $\mathbf{H} \cdot \mathbf{H}^1$ is not closed. Let us consider a regular foundation of $\mathbf{H} \cdot \mathbf{H}^1$ obtained from $\mathfrak{D}_1$ by eliminating the unnecessary applications of **Б**; that is, applications with outer component $\mathbf{H}^1$ and superfluous applications below. Replacing in it every closed sequence containing $\mathbf{H}^1$ (for example, a sequence $\mathbf{H}^1 \cdot \mathbf{H}^\beta$, where $\mathbf{H}^\beta$ is as in the preceding paragraph) by a sequence $\mathbf{H} \cdot \mathbf{H}^\beta$ of M_{ix} and then eliminating applications of Rule **Б** which become superfluous, we arrive at a foundation of favorability of the sequence $\mathbf{H}$, generally not containing $\mathbf{H}^1$. By the inductive hypothesis we have what is required ($\mathbf{H}^1$ can enter into the final foundation of $\mathbf{H}$ if M_{ix} was extended in the course of the reconstruction).

Lemma 18 is proved; in this connection the essential property of the reconstructions applied in the proof is that "good" occurrences in the tree of every sequence $\mathbf{H}^\gamma$ such that $\mathbf{H} \cdot \mathbf{H}^\gamma$ is $\{M^c\}$-derivable remain "good".

The proof of Theorem 6 is carried out first for constant trees according to the scheme of the proof of Lemma 9. In this connection there are included, of course, in the "proper part" the Π-ordered sequences derived in accordance with wΠ- and sup M-tactics. With regard to this change, we define $\Xi_{\mathfrak{D}}$ and $\mathbf{H}_{\mathfrak{D}}$ and assume that $\mathbf{H}$ is not $\{M^c\}$-derivable. We carry out the induction as in Lemma 9, but as inductive parameter we choose the number of elements of $\Xi_{\mathfrak{D}}$ occurring in $\mathfrak{D}$ "badly", in the sense of Lemma 18.

Case 1. There is a "badly" occurring element $\mathbf{H}^\gamma$ of $\Xi_{\mathfrak{D}}$ such that $\mathbf{H} \cdot \mathbf{H}^\gamma$ is $\{M^c\}$-derivable. We eliminate by Lemma 18 all its "bad" occurrences and we have a decrease in the inductive parameter.

Case 2. All elements described in Case 1 occur only as part of closed sequences. We choose from the "badly" occurring elements some element ξ (the first one according to Π) and we carry out the same reconstruction as in Lemma 9. For the applicability of the inductive hypothesis to a foundation with lowest sequence $\mathbf{H}^\xi$ it is essential that this sequence not be $\{M^c\}$-derivable. For both of the inductive transformations required in accordance with the proof of Lemma 9, it is essential that elements which (in accordance with Π) it is desirable to have to the left of the element ξ occur only in closed sequences (or in $\mathbf{H}_{\mathfrak{D}}$) and consequently do not violate the requirements of the wΠ-tactic. The reconstructions which are carried out yield as a result a tree satisfying the wΠ-tactic and the sup M-tactic and not containing supersequences of closed sequences as results of applications of **Б**.

The transition to nonconstant trees and the final proof of the theorem are carried out as in §3.6 (taking into account what was said in §4.2).

REMARK 7. Looking over the proof, we easily see that the necessity for permuting components of closed sequences arises only for sequences not supported by M. Therefore we can add to the conclusion of Theorem 6 the condition that all non-Π-ordered sequences occurring in the dc-foundation are closed sequences not supported by M.

BIBLIOGRAPHY

1. S. Ju. Maslov, *The inverse method for establishing deducibility for logical calculi*, Trudy Mat. Inst. Steklov. **98** (1968), 26–87 = Proc. Steklov Inst. Math. **98** (1968), 25–96. MR **40** #5416; **43** #4620.

2. J. A. Robinson, *A review of automatic theorem-proving*, Proc. Sympos. Appl. Math., vol. 19, Amer. Math. Soc., Providence, R. I., 1967, pp. 1–18. MR **39** #2540.

3. J. R. Slagle, *Automatic theorem-proving with renamable and semantic resolution*, J. Assoc. Comput. Mach. **14** (1967), 687–697. MR **38** #3093.

4. R. Kowalski and P. J. Hayes, *Semantic trees in automatic theorem proving*, Machine Intelligence, 4, American Elsevier, New York, 1969, pp. 87–101. MR **41** #1537.

5. N. K. Zamov and V. I. Šaronov, *A certain class of strategies used in theorem proving by the resolution method*, Zap. Naučn. Sem. Leningrad. Otdel. Mat. Inst. Steklov. (LOMI) **16** (1969), 54–64 = Sem. Math. V. A. Steklov Math. Inst. Leningrad **16** (1969), 26–31. MR **42** #2715.

6. S. Ju. Maslov, *The connection between tactics of the inverse method and the resolution method*, Zap. Naučn. Sem. Leningrad. Otdel. Mat. Inst. Steklov. (LOMI) **16** (1969), 137–146 = Sem. Math. V. A. Steklov Math. Inst. Leningrad **16** (1969), 69–73. MR **42** #1645.

7. ———, *Proof-search strategies for methods of the resolution type*, Machine Intelligence, 6, American Elsevier, New York, 1971, pp. 77–90.

8. D. G. Kuehner, *A note on the relation between resolution and Maslov's inverse method*, Machine Intelligence, 6, American Elsevier, New York, 1971, pp. 73–76.

9. S. Ju. Maslov, *Inference-seeking tactics that are based on a unification of the order of terms in a favorable set*, Zap. Naučn. Sem. Leningrad. Otdel. Mat. Inst. Steklov. (LOMI) **16** (1969), 126–136 = Sem. Math. V. A. Steklov Math. Inst. Leningrad **16** (1969), 64–68. MR **41** #8173.

10. G. Gentzen, *Untersuchungen über das logische Schliessen*, Math. Z. **39** (1934/35), 176–210, 405–431; Russian transl. in Mathematical Theory of Logical Inference, "Nauka", Moscow, 1967, pp. 9–74; English transl. in M. E. Szabo, *The collected papers of Gerhard Gentzen*, Studies in Logic and the Foundations of Math., North-Holland, Amsterdam, 1969, pp. 68–103. MR **36** #4957; **41** #6660.

11. S. C. Kleene, *Permutability of inferences in Gentzen's calculi LK and LJ. Two papers on the predicate calculus*, Mem. Amer. Math. Soc. No. 10 (1952); Russian transl. in Mathematical Theory of Logical Inference, "Nauka", Moscow, 1967, pp. 208–236. MR **14**, 439; **36** #4957.

12. S. Ju. Maslov, *An inverse method for establishing deducibility of nonprenex formulas of the predicate calculus*, Dokl. Akad. Nauk SSSR **172** (1967), 22–25 = Soviet Math. Dokl. **8** (1967), 16–19. MR **35** #19.

13. G. V. Davydov, S. Ju. Maslov, et al., *A machine algorithm for establishing deducibility on the basis of the inverse method*, Zap. Naučn. Sem. Leningrad. Otdel. Mat. Inst. Steklov. (LOMI) **16** (1969), 8–19 = Sem. Math. V. A. Steklov Math. Inst. Leningrad **16** (1969), 1–6. MR **41** #3278.

14. G. S. Ceĭtin, *The complexity of a deduction in the propositional predicate calculus*, Zap. Naučn. Sem. Leningrad. Otdel. Mat. Inst. Steklov. (LOMI) **8** (1968), 234–259 = Sem. Math. V. A. Steklov Math. Inst. Leningrad **8** (1968), 115–125. MR **43** #6098.

15. B. Dreben, P. Andrews and S. Aandreaa, *False lemmas in Herbrand*, Bull. Amer. Math. Soc. **69** (1963), 699–706. MR **27** #2401.

16. B. Dreben and S. Aandreaa, *Herbrand analyzing functions*, Bull. Amer. Math. Soc. **70** (1964), 697–698. MR **30** #1924.

17. G. E. Minc, *Herbrand's theorem*, Mathematical Theory of Logical Deduction, "Nauka", Moscow, 1967, pp. 311–350. (Russian) MR **36** #4958.

18. V. P. Orevkov, *Two unsolvable classes of formulas of the classical predicate calculus*, Zap. Naučn. Sem. Leningrad. Otdel. Mat. Inst. Steklov. (LOMI) **8** (1968), 202–210 = Sem. Math. V. A. Steklov Math. Inst. Leningrad **8** (1968), 98–102. MR **42** #5776.

19. S. Ju. Maslov and V. P. Orevkov, *Decidable classes reducing to a one-quantifier class*, Trudy Mat. Inst. Steklov. **121** (1972), 57–66 = Proc. Steklov Inst. Math. **121** (1972), 61–72.

20. G. V. Davydov, *A method of establishing deducibility in the classical predicate calculus*, Zap. Naučn. Sem. Leningrad. Otdel. Mat. Inst. Steklov. (LOMI) **4** (1967), 8–17 = Sem. Math. V. A. Steklov Math. Inst. Leningrad **4** (1967), 1–4. MR **39** #69.

21. S. Kanger, *A simplified proof method for elementary logic*, Computer Programming and Formal Systems, North-Holland, Amsterdam, 1963, pp. 87–94; Russian transl. in Mathematical Theory of Logical Inference, "Nauka", Moscow, 1967, pp. 200–209. MR **27** #1350.

22. S. Ju. Maslov, *An invertible sequential version of the constructive predicate calculus*, Zap. Naučn. Sem. Leningrad. Otdel. Mat. Inst. Steklov. (LOMI) **4** (1967), 96–111 = Sem. Math. V. A. Steklov Math. Inst. Leningrad **4** (1967), 36–42. MR **39** #6736.

23. N. A. Šanin, G. V. Davydov, et al., *An algorithm for a machine scan of a natural logical deduction in a propositional calculus*, Academy of Sciences of the USSR, V. A. Steklov. Mathematical Institute, Leningrad Branch, "Nauka", Moscow, 1965. (Russian) MR **33** #5405.

24. V. P. Orevkov, *On Glivenko sequent classes*, Trudy Mat. Inst. Steklov. **98** (1968), 131–154 = Proc. Steklov Inst. Math. **98** (1968), 147–176. MR **39** #2606; **43** #4620.

Translated by E. MENDELSON

Trudy Mat. Inst. Steklov.
121 (1972)

Proc. Steklov Inst. Math.
121 (1972)

DECIDABLE CLASSES REDUCING TO A ONE-QUANTIFIER CLASS

S. Ju. MASLOV AND V. P. OREVKOV

UDC 51.01:164

ABSTRACT. Decidability with respect to provability in the classical predicate calculus with function symbols (without equality) is considered; for the concept of F-prefix and the type of formulas under consideration cf. MR 35 #25. $F \in \mathbf{N}_2$ if and only if every term of F contains not more than one variable and every F-prefix whose length is greater than 1 terminates with a $\forall$-quantifier. The decidability of the class of arbitrary disjunctions of $\mathbf{N}_2$ is proved. (Both conditions imposed on an F of $\mathbf{N}_2$ are necessary.) A number of well-known decidable classes, in particular the class $\mathbf{N}_1$ of formulas containing not more than one $\exists$-quantifier, get into $\mathbf{N}_2$ (after Skolemization). The decidable class obtained from $\mathbf{N}_1$ just as $\mathbf{K}'$ from $\mathbf{K}$ (MR 40 #5416) is considered. The basic proof device is the inverse method (utilizing miniscopizations and simulation of favorable sets).

Bibliography: 11 items.

1. Let us denote by $\mathbf{C}$ the classical predicate calculus with function symbols (without equality). All formulas considered below are implicitly in the language of this calculus.

The purpose of this article is proving the decidability with respect to provability of the class $\mathbf{N}_1$ of formulas containing only one occurrence of the existential quantifier (cf. [1]), and of some other classes reducible to $\mathbf{N}_1$. The proofs have a syntactical character. (The concepts of model and general validity are not utilized.) *Miniscopization* (diminishing the scopes of quantifiers) and the *inverse method* of establishing provability (cf. [2]) are the basic technical apparatus.

The miniscopization procedures are based on widely known equivalences (cf., for example, assertions *87–*92 of [3]). A brief description of the inverse method is cited in §2 (for more detail, see Chapter II of [2]). Only the simplest of the instances worked out in Chapter II of [2] will be utilized. (Thus, for example: 1) the proof of Theorem 1 is reminiscent of the proof of the decidability of class $\mathbf{K}$ but turns out to be substantially simpler since during the basic induction step free one-element sets emerge and not the comparatively complex multi-element decomposable sets of [2], §10.2; 2) the simplest form of the method of simulating favorable sets, simulating

---

AMS (MOS) subject classifications (1970). Primary 02B10, 02G05.

closed sets by closed, is utilized in the proof of the lemma). Under further applications of the apparatus set forth this makes it possible to hope to obtain results essentially stronger than that of Theorem 3. (Although the class whose decidability is established by this theorem is fairly broad, it includes in particular all (see [4]) decidable classes of prenex formulas of the calculus **C** which can be formulated in terms of a prefix and a signature.)

2. Convenient for establishing the decidability of concretizations of the inverse method is a concretization in all respects analogous to that offered in §8 of [2] (and so impractical from the point of view of constructing machine algorithms), except for the fact that systems of relations are prescribed not with the help of the concept of a basis but, as was done for **C** in [5], [6], with the help of the concept of a bundle.

Let us assume that the sequents examined are of the form

$$\rightarrow \mathrm{K}_1 z_1 \mathrm{K}_2 z_2 \ldots \mathrm{K}_m z_m \left(\mathop{\&}_{i=1}^{\delta} D_i \right), \tag{1}$$

where $\mathrm{K}_1 z_1, \cdots, \mathrm{K}_m z_m$ are quantifier complexes and $D_1, \cdots, D_\delta$ are prime disjunctions, i.e. disjunctions of elementary formulas (atomic formulas and negations of atomic formulas). Let us denote the succedent of sequent (1) by F. Without loss of generality, let us assume that $\mathrm{K}_1 = \exists$ and F does not contain free variables (but may contain zero-place function symbols). Let C be a list of natural numbers $\alpha_1, \cdots, \alpha_h$, $h \geqslant 0$, not exceeding δ; we regard any expression of the form z_μ^k, where $1 \leqslant \mu \leqslant m$ and $1 \leqslant k \leqslant h$, as an (F, C)-atom. Let us call an *F-term* any term in whose make-up function symbols of F and any variables occur. (Exactly they play the role of the expressions $\lambda, \lambda', \lambda'', \cdots$ in the definition of F-term in [5].) Let us call a list of *relations*, i.e. of expressions of the form $A \approx T$, an (F, C)-*bundle*. (Here A is an (F, C)-atom having a unique occurrence in the bundle, and T is an F-term.) Let us call only the variables in the make-up of F-terms, the *variables of the bundle*, and stipulate that $z_1, \cdots, z_m$ are not among them. Let us establish a total ordering of (F, C)-atoms: z_μ^k precedes z_ν^l if either $k < l$ or $k = l$ but $\mu < \nu$. Let us say that an (F, C)-bundle S is *admissible* if, for any k and μ $(1 \leqslant k \leqslant h; 1 \leqslant \mu \leqslant m)$ such that $\mathrm{K}_\mu = \forall$ and $z_\mu^k \approx T$ occurs in S, T is a variable and one of the following conditions is fulfilled:

1) For any relation of the form $A \approx T'$ in S, where A precedes z_μ^k, T does not occur in T'.

2) There is a $k' < k$ such that the relations $z_1^{k'} \approx T_1, \cdots, z_\mu^{k'} \approx T_\mu, z_1^k \approx T_1, \cdots, z_\mu^k \approx T_\mu$ occur in S, where $T_1, \cdots, T_\mu$ are F-terms ($T_\mu = T$).(1)

The pair $[\mathrm{C}; S]$, where S is an admissible (F, C)-bundle, is called an *F-set.* The set $[\mathrm{C}; S]$ is called a *subset* of the set $[\alpha'_1, \cdots, \alpha'_h; S]$ if for some numbers $j_1, \cdots, j_h$ $(1 \leqslant j_1 < \cdots < j_h \leqslant h')$: 1) $\alpha'_{j_i} = \alpha_i$ $(i = 1, \cdots, h)$ and 2) for the complete list $u_1, \cdots, u_s$ of variables of S there is a list of F-terms $T_1, \cdots, T_s$ such that for every relation $z_\mu^k \approx T$ in S the relation(2) $z_\mu^{jk} \approx [T]_{T_1 \ldots T_s}^{u_1 \ldots u_s}$ is in S. Let us call the formula $W_1 \vee \cdots \vee W_h$, where W_i is obtained from D_{α_i} by substituting the term T in place of the variable z_μ, the formula *corresponding to the set* $[\mathrm{C}; S]$ if $z_\mu^k \approx T^k$ is contained in S and otherwise some new variable (each time its own and different from the variables of S). A set is called *closed* if the disjunction corresponding to it is tautological and "+"-*closed* if, in addition, none of its subsets is closed.

Now let us define inductively the concept of a *favorable set.*

Rule A. Any closed set is considered favorable.

Rule Б. Let $[\mathrm{C}_1, 1; S_1], \cdots, [\mathrm{C}_\delta, \delta; S_\delta]$ be favorable sets. The set $[\mathrm{C}; S]$ is considered favorable if there are relations of the form $z_{\mu_1}^{h+1} \approx T_1, \cdots, z_{\mu_f}^{h+1} \approx T_f$ $(\mu_1, \cdots, \mu_f$ all distinct) after adding which to S we obtain an admissible bundle S' such that for all i, $1 \leqslant i \leqslant \delta$, $[\mathrm{C}_i, i; S_i]$ is a subset of $[\mathrm{C}, i; S']$.

For the above-outlined formulation of the inverse method completeness is proved in the usual way (*F is provable in* C *if and only if the empty F-set is favorable*), as well as other results (cf., for example, §§8 and 9 of [2]). Let us note the following obvious result: *if an F-set of the form* $[\alpha;]$, i.e. a free one-element set, *is favorable, F is equivalent to the formula obtained by eliminating the disjunction D_α from F. (If, moreover,* $\alpha = \delta = 1$, *then F is provable.)*

Let us define the *height of a term* in the following way. The number 0 is taken as the height of variables and 0-place function symbols. If the heights of the terms $t_1, \cdots, t_k$ are $n_1, \cdots, n_k$ and f is a k-place $(k > 0)$ function symbol, the number $\max(n_1, \cdots, n_k) + 1$ is taken as the height of the term $f(t_1, \cdots, t_k)$. Let us denote by r_F the maximal height of the terms occurring in F. Let us call the

---

(1) Let us note that in the definition of the admissibility of a bundle we have not required fulfillment of conditions corresponding to conditions 2) and 3) of the definition of a permissible basis ([2], p. 58 = 61). This is connected with the fact that when including inferences ([2], §7.2; the unnegated Russian term is правильный вывод, which was translated as "correct deduction" in the English version of [2]) in the discussion fulfillment of the mentioned conditions is not necessary to comprehend the basic interpretation of a bundle as a system of relations between the various substitutions which may be encountered in a search for deduction trees.

(2) The expression $[\mathrm{A}]_t^x$ denotes the result of substituting the term t in place of all free occurrences of the variable x in the formula or term A.

terms whose height does not exceed r_F and in which only function symbols(3) of F and variables of the formula F can occur, *F-allowed*. Let us call an elementary formula an *F-allowed* formula if its predicate letter is encountered in F and its terms are F-allowed terms.

3. THEOREM 1. *The problem of provability in* **C** *is solvable for the class* $\mathbf{N}_1$.(4)

PROOF. Let F be a formula of the class $\mathbf{N}_1$ of the form

$$\exists x \left(\mathop{\&}_{i=1}^{\delta} D_i \right);$$

any formula of $\mathbf{N}_1$ can be reduced to this form by eliminating occurrences of universal quantifiers through the introduction of new function symbols (see, for example, [8], p. 331). Let us assume that the formula F is saturated with respect to all possible F-allowed formulas E. (Saturation consists of replacing a disjunction D not containing the atomic formula E with the formula $(D \vee E) \& (D \vee \neg E)$; cf. [2], p. 66 = 70.) The class of such formulas with $\delta = 1$ is obviously decidable. Let us carry out induction on δ.

If F is provable, a complete weakly minus-normal tree(5) of its deduction can be constructed in each branch of which: 1) a quantifier is not taken down twice with the same substitution for x; 2) substitution of any subterm of T for x is encountered below substitution of T for x; 3) substitutions of x of terms of height greater than r_F are higher than all substitutions in which the height of the substituted term does not exceed r_F. If such a deduction can be constructed with substitution of terms not exceeding r_F in height, it can be established in a finite number of steps. Let F be provable, and let

$$\frac{\to \Gamma,\ \bar{D}_1,\ \ldots,\ \Gamma,\ \bar{D}_\delta}{\to \Gamma}$$

be some uppermost application of the rule $(\to \&)$, i.e. all $\to \Gamma, D_i$ are axioms, while $\to \Gamma$ is not an axiom; when obtaining $\bar{D}_\alpha$ from D_α $(\alpha = 1, \cdots, \delta)$, let a term T whose height is greater than r_F have been substituted for x.

Let us suppose that the term T is not essential, i.e., when replacing it by any other, axioms (of the form $\to \Gamma,\ [D_i]^x_{T'}$) are also obtained. Then, replacing T by some term which was substitued lower on the branch, we obtain the result that now

(3) We recall that free variables are considered zero-place function symbols.

(4) This result was announced for the first time in 1965 at the Second All-Union Symposium on Cybernetics in Tbilisi [1]. Another proof of this fact is cited in [7]. (There the solvability of the problem of finite realizability is also established for a corresponding class of formulas.)

(5) For the definition of the completeness of a deduction, cf. [2], §7.2; for the definition of *weak-minus-normality*, cf. [6], §6.3.

$\to \Gamma$ is an axiom. Consequently there are α and a formula E of D_α which does not contain x fictitiously and is in the contradictory pair ensuring the axiomality of the sequent $\to \Gamma, \bar{D}_\alpha$. The height of the term T guarantees us that the second element of this contradictory pair is obtained from some lower lying formula E' containing x. Let T_0 be the maximal term for which an F-allowed term t, not a variable, exists such that $T = [t]^x_{T_0}$. The possibility of such a representation (not necessarily with maximal T_0) is elicited from E'; uniqueness follows from the fact that the height of T does not exceed r_F.

Let us find the place in the branch under consideration at which some formula $\bar{D}_\beta$ is obtained by substituting the term T_0 for x in D_β. A formula contrary to any of the D_α $(1 \leqslant \alpha \leqslant \delta)$ necessarily lies in Γ and consequently (because of saturation) occurs in $\bar{D}_\beta$. This holds for all α (otherwise $\to \Gamma$ would be an axiom). Thus we have the closed sets $[\beta, \alpha; x^1 \approx u; x^2 \approx [t]^x_u]$ $(1 \leqslant \alpha \leqslant \delta)$.(6) As the result of a single application of Rule Б we arrive at the free set $[\beta;]$. Consequently for any F provable only with the help of terms of height greater than r_F can we diminish δ with the help of $[\beta;]$. The lemma is proved.

4. Reviewing the miniscopization applied when reducing $\mathbf{K}'$ to $\mathbf{K}$ ([2], § 10.3), it is easy to establish that the absence of function symbols in the language is not utilized at all in the proof. Moreover, the miniscopization applied does not increase the length of F-prefixes. In particular, the maximal length of an F-prefix of the form $\exists x_1 \cdots \exists x_k$ in the original formula (of class $\mathbf{K}'$) does not increase. Therefore the *class* $\mathbf{N}_1'$ obtained from the class $\mathbf{K}'$ by adding function symbols to the language and replacing requirement 4) by the more rigid, "each group Γ_i contains not more than one $\exists$-quantifier" is reduced to $\mathbf{N}_1$ by the same transformations as in [2].

Hence the following holds.

THEOREM 2. *The class* $\mathbf{N}_1'$ *is decidable with respect to provability.*

It is clear that the class $\mathbf{N}_1$ is contained in $\mathbf{N}_1'$. Let us observe also that in $\mathbf{N}_1'$ are contained the class $\mathbf{M}$ of [9] (the class of formulas F in each F-prefix of which all quantifiers, except perhaps only the last, are universal quantifiers) and consequently the class Φ (*All*, (∞), (∞)) of [4] (the class of prenex formulas with any prefix which contain only one-place predicate variables and function symbols).

A different formulation of the class $\mathbf{N}_1'$ is convenient in certain cases. Let F be a formula of the calculus $\mathbf{C}$.(7) Let us say that the quantifier complex $Q_1 x_1$

---

(6) It is assumed here that there are no tautologies among $D_1, \cdots, D_\delta$, since the corresponding closed sets would turn out to be one-element free ones.

(7) We recall that we consider only formulas which: 1) do not contain logical symbols besides $\forall$, $\exists$, &, $\vee$ and $\neg$; 2) contain $\neg$ only in the make-up of elementary formulas; and 3) do not have distinct occurrences of quantifier complexes whose characteristic variables coincide with each other or with a free variable of the formula.

completely dominates the complex Q_2x_2 in the formula F if it is possible to construct an F-prefix in which Q_1x_1 precedes Q_2x_2. Let us say that Q_1x_1 *strongly dominates* Q_2x_2 in the formula F if Q_2x_2 is in the scope of Q_1x_1 and one of the following cases is implemented:

1) Q_1x_1 completely dominates Q_2x_2 in F.

2) It is possible to construct a complex which strongly dominates Q_1x_1 in F and which completely dominates Q_2x_2.

3) Q_2x_2 is a universal complex and it is possible to construct an F-prefix in which only universal complexes, Q_2x_2 and any complex which strongly dominates Q_1x_1 occur.

Let us say that the formula F *belongs to the class* $\mathbf{N}_1'$ if, whatever the complexes $\exists u$ and $\exists v$, they strongly dominate the same complex in the formula F if and only if u coincides with v.

5. Let us say that a formula F (possibly nonclosed) *belongs to the class* $\mathbf{N}_2$ if the following conditions are fulfilled:

1) Every term occurring in F contains not more than one variable.

2) Every F-prefix whose length is greater than 1 terminates with a universal quantifier.

$\mathbf{N}_1$ (after Skolemization) and the class of formulas containing only one-place predicate and function variables lie in $\mathbf{N}_2$ (cf. [4]).

THEOREM 3. *The class* $\mathbf{N}_2$ *(and consequently the class of arbitrary disjunctions of formulas of* $\mathbf{N}_2$*) is decidable with respect to provability.*

The proof of the theorem is divided into three steps.

5.1. Let us observe that reduction to prenex form does not take a formula out of $\mathbf{N}_2$.

LEMMA. *Let* F *be a formula of* $\mathbf{N}_2$, *of the form* $\mathrm{P}M$ (M *being quantifier-free). It is possible to construct deductively a formula* $\widetilde{F}$ *of* $\mathbf{N}_2$ *which is equivalent to it and of the form* $\mathrm{P}(\widetilde{M})$, *where besides elementary formulas with an* $\widetilde{F}$*-prefix whose length does not exceed* 1, *only formulas of the form*

$$\forall \tilde{y}\,(Q\,(\tilde{y},\ x_1, \ldots, x_s) \vee \neg\, Q\,(\tilde{y},\ T_1, \ldots, T_s)) \tag{2}$$

and variables $x_1, \cdots, x_s$ *bound in* $\widetilde{F}$ *by the quantifier* $\exists$ *occur in the make-up of* $\widetilde{M}$.

PROOF. Let us reduce M to conjunctive form and implement saturation with respect to all F-allowed formulas. As the result we arrive at the formula

$$\mathrm{P}\left(\mathop{\&}_{i=1}^{\delta} D_i\right), \tag{3}$$

which is equivalent to the original and is in $\mathbf{N}_2$.

Let us denote by $\mathfrak{M}^1$ the collection of one-element "+"-closed (3)-sets. Let $[\alpha_1, \cdots, \alpha_h; \beta_1, \beta_2; S]$ be some "+"-closed multi-element set ($h \geqslant 0$), and let $\bar{D}_{\alpha_1} \vee \cdots \vee \bar{D}_{\alpha_h} \vee \bar{D}_{\beta_1} \vee \bar{D}_{\beta_2}$ be the formula corresponding to this set.

Let us consider the contradictory pair $E, \neg E$ ensuring the closedness of this set. Without loss of generality it is possible to assume that E occurs in $\bar{D}_{\beta_1}$ and $\neg E$, in $\bar{D}_{\beta_2}$. (The first h members are needed only for the admissibility of S; cf. [2], §9.4.) Let E_1 and E_2 be elementary formulas of D_{β_1} and D_{β_2}, respectively, by means of substitutions into which E and $\neg E$ are obtained. For one of these formulas, say for $\neg E_2$, let the (3)-prefix be of the form $\exists x$. Hence there are F-allowed terms $T_1, \cdots, T_\epsilon$, a term T and a predicate variable R for which $E = R([T_1]^x_T, \cdots, [T_\epsilon]^x_T)$ and $E_2 = R(T_1, \cdots, T_\epsilon)$.

Let us assume that the (3)-prefix of the formula E_1 is of more than unit length. Let us prove that then it has the form $\exists x_1 \cdots \exists x_\tau \forall y_0$ and when obtaining E substitution of F-allowed constant terms for $x_1, \cdots, x_\tau$ was realized in E_1.

Indeed, exactly the universal $\forall y_0$ will be the terminal complex of the prefix under consideration. Under the contemplated substitution, in place of y_0 is put some variable t which necessarily occurs in the above-mentioned term T. On the other hand, because of the admissibility of S, t cannot occur in the make-up of any term substituted for the prefix variables preceding y_0. Consequently t is not in the interior of these terms, and the above-mentioned relationship between E and E_2 guarantees that these terms are constant and F-allowable. This in turn is possible only when all variables preceding y_0 in the (3)-prefix under consideration are bound by an existential quantifier.

Hence $R([T_1]^x_T, \cdots, [T_\epsilon]^x_T) = [E_1]^{x_1, \ldots, x_\tau, y_0}_{\widetilde{T}_1, \ldots, \widetilde{T}_\tau, t}$, $\widetilde{T}_1, \cdots, \widetilde{T}_\tau$ being F-allowed and constant. Then the maximal height of the terms occurring in this atomic formula does not exceed r_F. ($[T_i]^x_T$ either equals T_i or is obtained from some term occurring in E_1 by substituting t for y_0.) Consequently the formula E_1', equal to $[E_1]^{x_1, \ldots, x_\tau}_{\widetilde{T}_1, \ldots, \widetilde{T}_\tau}$ is F-allowed, i.e. it must (by virtue of saturatedness) occur in D_{β_1}. Then it occurs in D_{β_1} without negation. (Otherwise the set under consideration would turn out to be a superset of the one-element closed set $[\beta_1; x_1^1 \approx \widetilde{T}_1, \cdots, x_\tau^1 \approx \widetilde{T}_\tau]$.) Thus to obtain a closed set instead of E_1 it is possible to avail ourselves of the formula E_1', which has a (3)-prefix of length 1. (By the way, the closed set under consideration would turn out to be the following: $[\beta_1, \beta_2; y_0^1 \approx t, x^1 \approx T]$.)

Now let the (3)-prefixes both of E_1 and of $\neg E_2$ terminate with a universal complex. Then, as is easy to understand, these complexes are one and the same (let us denote it by $\forall y_0$), and for any variable z not to the right of y_0 in the prefix P there are the relations $z^{h+1} \approx T_z$ and $z^{h+2} \approx T_z$ in S. By virtue of saturatedness, E_2 occurs in D_{β_1}, without negation. (Otherwise we would have a one-element closed subset $[\beta_1; S']$.) Consequently $\beta_1 \neq \beta_2$, and the set under consideration has the form

$$[\beta_1, \beta_2;\ z_1^1 \approx u_1,\ \ldots,\ z_\rho^1 \approx u_\rho,\ y_0^1 \approx u,\ z_1^2 \approx u_1,\ \ldots,\ z_\rho^2 \approx u_\rho,\ y_0^2 \approx u], \qquad (4)$$

where $z_1, \cdots, z_\rho$ is the complete list of variables preceding y_0 in P and $u, u_1, \cdots, u_\rho$ are distinct variables. As a matter of fact, the relations written out are just those which are necessary for the closedness of the disjunction $\bar{D}_{\beta_1} \vee \bar{D}_{\beta_2}$ with respect to the contradictory pair $\bar{E}_2, \neg \bar{E}_2$, where $\bar{E}_2 = [E_2]^{z_1, \ldots, z_\rho, y_0}_{u_1, \ldots, u_\rho, u}$.

Hence "+"-closed (3)-sets either are in $\mathfrak{M}^1$ or have the form (4) or are obtained with the help of elementary formulas with an empty or unary (3)-prefix. Now the problem consists of throwing out all elementary formulas with a longer prefix and instead of them writing in others (which make it possible to obtain $\widetilde{F}$) so that the collection of all possible "+"-closed sets does not change; i.e. we must "simulate" sets of $\mathfrak{M}^1$ and sets of the form (4).

For each set of the form (4) let us introduce a new one-place predicate variable R' (for each its own) and write the formula $R'(y_0)$ into D_{β_1} and the formula $\neg R'(y_0)$ into D_{β_2}. Now let us simulate an arbitrary set $[\alpha; S]$ of $\mathfrak{M}^1$. Let u be some variable of the bundle S;([8]) let us examine the term T of the least height, containing u, and occurring in S in some relation $A \approx T$. If the value of the atom A occurs in the contradictory pair ensuring the closedness of the set $[\alpha; T]$, then for any relation $A' \approx T'$ in which T' contains u and occurs in that contradictory pair, we have that T occurs in T'. The "+"-closedness implies that $T = u$. (Otherwise, having replaced all occurrences of T in S with u, we would obtain a proper subset.) Therefore our set can be written as

$$[\alpha;\ z_1^1 \approx u_1,\ \ldots,\ z_f^1 \approx u_f,\quad x_1^1 \approx T_1,\ \ldots,\ x_s^1 \approx T_s],$$

where $u_1, \cdots, u_f$ is the complete list, without repetitions, of the variables of S, each of these variables occurring in at least one of the terms $T_1, \cdots, T_s$ and not more than one of the variables $u_1, \cdots, u_f$ occurring in each of the terms $T_1, \cdots, T_s$. Let us bear in mind that with the possibility of including one of the atoms A and A' in the aggregate of atoms $z_1, \cdots, z_f$, the one whose variable is more to the left in the prefix P is chosen. This guarantees the impossibility of a universal variable's getting into the make-up of the atoms $x_1^1, \cdots, x_s^1$. Let us choose some new variables, an object one $\widetilde{y}$ and an $(s+1)$-place predicate one Q. Let us append $\forall \widetilde{y}$ to the prefix P on the right and add the two elementary formulas

$$Q(\tilde{y},\ x_1,\ \ldots,\ x_s) \quad \text{and} \quad [\neg Q(\tilde{y},\ T_1, \ldots, T_s)]^{u_1, \ldots, u_f}_{z_1, \ldots, z_f}.$$

to D_{α_s}. As the result of this kind of replacement for all sets in $\mathfrak{M}^1$ and for all of the above-mentioned sets (4), we arrive at the formula F'. If all old elementary formulas, i.e. those occurring in (3), whose F'-prefixes are greater than unity are thrown

([8]) Let us recall that only variables in the make-up of F-terms are called variables of the bundle.

out of this formula, we obtain the formula F'' with the same collection of "+"-closed sets.

The proof of the lemma is completed by applying well-known equivalences resulting in diminution of the scope (miniscopization) of the quantifiers of formula F'' which do not occur in the prefix P, i.e. of the quantifiers $\forall$.

5.2. Let $\widetilde{F}$ be an arbitrary formula. Let us call the following its *primitive blocks*:

1) elementary formulas with $\widetilde{F}$-prefix of the form $\exists x$;

2) formulas of the form $\forall y_1 \cdots y_1 J$, where J is quantifier-free and contains only elementary formulas without existential variables;

3) formulas of the form $\exists u_1 \cdots u_l(\&_1^e B_i)$, where each of the B_i either is elementary and has as $\widetilde{F}$-prefix any of the complexes $\exists u_1, \cdots, \exists u_l$, or is a formula of the form (2), none of the variables $u_1, \cdots, u_l$ being among $x_1, \cdots, x_s$.

Let $\widetilde{F}$ be of the form PM, where M is made up of primitive blocks with the help of the symbols & and $\vee$. (The formula obtained in accordance with the lemma is such a one.) Let us implement miniscopization in combination with certain transformations of the primitive blocks which preserve deductive equivalence. Let us show by induction on the length of P that $\widetilde{F}$ reduces to a conjunction of the primitive blocks not containing formulas of the form (2) which is equivalent to it.

Let us describe the admissible transformations of blocks of the form 3). Let us consider a weeded(9) deduction of the formula $\widetilde{F}$ and some application of the rule $(\rightarrow \exists)$ with principal formula $\exists u_1 \cdots u_l(\&_1^e \overline{B}_i)$, where, for example, $\overline{B}_1$ is obtained from formula (2) as the result of some lower lying applications of $(\rightarrow \exists)$. Then a formula of the form $\forall \widetilde{y}(E \vee \neg E)$ necessarily emerges in $\overline{B}_1$; otherwise the application under consideration would be weeded. (Because of the presence of the variable $\widetilde{y}$, the formula $\overline{B}_1$ proves to be not superfluous only if it itself gives the contradictory pair.) This permits making the following inferences.

1) A primitive block containing several formulas of the form (2) can be replaced, preserving deductive equivalence, by a block with one such formula (if the one-element closed sets corresponding to the original formulas (2) are consistent) or by some standard falsity $p \& \neg p$ (in the presence of inconsistency of the corresponding closed sets and, consequently, of the impossibility of finding a substitution under which all formulas of type (2) of the original block turn into tautologies).

2) Let us denote (2) by **q** and the formula

$$\forall \widetilde{y}\,(Q'(\widetilde{y}, x_2, \ldots, x_s) \vee \neg Q'(\widetilde{y}, T_2, \ldots, T_s))$$

by **q**$'$. Preserving deductive equivalence, it is possible to replace a formula of the form

(9) See §2 of [10] (cf. Lemma 4 of [11]).

$$\exists x_1 (K' \& \exists u_1 \ldots u_l (\mathbf{q} \& K'')),$$

where K' is a conjunction of elementary formulas with the single variable x_1 and K'' is a prime conjunction (of a primitive block), by the formula

$$\exists u_1 \ldots u_l (\mathbf{q}' \& \ [K']^{x_1}_{T_1} \& K''),$$

if a variable of the list $u_1, \cdots, u_l$ occurs in T_1, and by the formula

$$([K']^{x_1}_{T_1} \& \exists u_1 \ldots u_l (\mathbf{q}' \& K''))$$

otherwise. It is borne in mind that Q' is a new s-place predicate variable; in the case $s = 1$ we replace the formula $\forall \tilde{y} (Q'(\tilde{y}) \vee \neg Q'(\tilde{y}))$ by the standard truth $p \vee \neg p$.

Let us consider a formula of the form $\bar{\mathrm{P}} \exists \bar{x} (K_1 \vee \cdots \vee K_\alpha)$, where $\bar{\mathrm{P}}$ is a quantifier prefix and $K_1, \cdots, K_\alpha$ are conjunctions of primitive blocks. For each K_i let us join the primitive blocks of type 3) which contain $\bar{x}$ into one, let us replace it by a falsity or a block with one formula of the form (2), and let us replace $\bar{x}$ by the variables $\bar{x}_1, \cdots, \bar{x}_{\alpha'}, \alpha' \leqslant \alpha$. (For definiteness we assume that just the last $\alpha - \alpha'$ conjunctions turn out not to contain $\bar{x}$.) As the result, we obtain the formula

$$\bar{\mathrm{P}} ((K_1' \& \exists \bar{x}_1 (K_1'' \& \Pi_1)) \vee \ldots \vee (K_{\alpha'}' \& \exists \bar{x}_{\alpha'} (K_{\alpha'}'' \& \Pi_{\alpha'})) \vee K_{\alpha'+1}' \vee \ldots \vee K_\alpha'), \quad (5)$$

where $K_{\alpha'+1}' \cdots K_\alpha'$ do not contain $\bar{x}$ and for each i, $1 \leqslant i \leqslant \alpha'$, K_i' does not contain $\bar{x}_i$, K_i'' is a conjunction of elementary formulas with the (5)-prefix $\exists \bar{x}_i$ and Π_i is a primitive block containing one formula of type (2).

Let us consider the formula $\exists \bar{x}_1 (K_1'' \& \Pi_1)$. Having used the above-described transformation of formulas of this kind, we either bring $\bar{x}_1$ into the make-up of the quantifiers of the primitive block (if a formula of type (2) in Π_1 does not contain $\bar{x}_1$) or in general get rid of that quantifier. The same can be done for $\bar{x}_2, \cdots, \bar{x}_{\alpha'}$ also.

Let us consider a formula of the form $\bar{\mathrm{P}} \forall \bar{y} J$, where J consists of primitive blocks. Let us consider the block $\exists u_1 \cdots u_l K$ containing a formula of type (2), and let us assume that $\bar{y}$ occurs in T_1. Then the only substitution transforming (2) into a tautology is impossible since x_1 precedes the variable $\bar{y}$ in $\bar{\mathrm{P}} \forall \bar{y}$. (By the definition of primitive block, x_1 is not among $u_1, \cdots, u_l$.) Therefore the block under consideration can be replaced by a falsity. Thus we may assume that there are no occurrences of the variable $\bar{y}$ among the blocks of form 3). Then the usual miniscopization permits bringing $\bar{y}$ into the make-up of blocks of form 2).

Application of the indicated transformations results in the complete removal of the outer prefix from the formula. As is easy to understand, all formulas of type (2) in the last analysis turn out to be replaced by a falsity or a truth. Q.E.D.

5.3. It remains to consider the formulas in $\mathbf{N}_2$ of the form

$$M_1 \vee \forall y_1 \ldots y_l M_2 \vee \exists x_1 \ldots x_m M_3 \tag{6}$$

(M_1, M_2 and M_3 are quantifier free). In view of the absence of elementary formulas containing more than one variable, we can convert (6) into a deductively equivalent formula of the form $\exists xM$ (M is quantifier free). Hence the proof of the theorem is completed.

6. Both the removal of the first, and the removal of the second, condition imposed on the class $\mathbf{N}_2$ result in reduction classes. This is obvious for the second condition. (Any formula of a pure predicate calculus satisfies the first condition.) Let us prove that the class of formulas F all of whose F-prefixes are empty or unary or which terminate with the quantifier $\forall$ is a reduction class.

It is obvious that a reduction class for $\mathbf{C}$ is the class of formulas of the form $\exists x_1 x_2 M$, where M is quantifier free and contains one one-place predicate variable R and some number not greater than that of the two-place function variables. (For each predicate variable, for example for the two-place variable P, its own function f_P can be introduced after replacing a formula of the form $P(T_1, T_2)$ by the formula $R(f_P(T_1, T_2))$.) Let F be a formula of the described class in whose make-up there is the atomic formula $R(T)$. Let us introduce a two-place predicate variable Q, replace $R(T)$ everywhere in F by $R(x_3)$ (we denote the matrix of the resulting formula by M'), and construct the formula

$$\exists x_1 x_2 x_3 (M' \,\&\, \forall \bar{y} (Q(\bar{y}, x_3) \vee \neg Q(\bar{y}, T))).$$

Taking into account the contents of §5.2, it is easy to prove the deductive equivalence of this formula and the formula F. By applying this step successively we can get rid of all occurrences in F of terms containing more than one variable. As a result only elementary formulas with empty or unary F'-prefixes will be contained in what remains of the matrix M. (F' is the notation for the formula obtained as the result of the described transformations.) On the other hand, the F'-prefixes of all formulas of form (2) which are introduced will terminate with the quantifier $\forall \bar{y}$.

Q.E.D. It is not difficult to develop further regulation of the described reduction class. For example, it is possible to add the requirement that only two atomic formulas in F have an F-prefix of greater than unary length, where $r_F = 1$. (The height of a variable is zero.) Nevertheless, the first requirement imposed on the class $\mathbf{N}_2$ can be relaxed. *The proof of* §5 *holds, with insignificant changes, for a class with the following formulation of the first condition*: *whatever the term* T *occurring in* F, *if more than one variable occurs in* T, *all variables occurring in* T *are bound in* F *by a universal quantifier.*

As regards changes in the proof, let us note only that in the formulation of the lemma (and in the proofs of 5.1 and 5.2) instead of $\widetilde{F}$-prefixes whose length does not exceed 1, we must speak of $\widetilde{F}$-prefixes of the forms $\exists x$ and $\forall y_1 \cdots \forall y_l$ ($l \geqslant 0$).

BIBLIOGRAPHY

1. V. P. Orevkov, *One solvable class of formulas of the classical predicate calculus with functional symbols*, II Sympos. on Cybernetics, Abstracts of Reports, Tbilisi, 1965, p. 176. (Russian)

2. S. Ju. Maslov, *The inverse method for establishing deducibility for logical calculi*, Trudy Mat. Inst. Steklov. **98** (1968), 26–87 = Proc. Steklov Inst. Math. **98** (1968), 25–96. MR **40** #5416; **43** #4620.

3. S. C. Kleene, *Introduction to metamathematics*, Van Nostrand, Princeton, N. J., 1952; Russian transl., IL, Moscow, 1957. MR **14**, 525; **19**, 2.

4. Ju. Š. Gurevič, *The problem of reduction for the logic of predicates and operations*, Algebra i Logika **8** (1969), 284–308 = Algebra and Logic **8** (1969), 160–174. MR **41** #8205.

5. S. Ju. Maslov, *Inference-seeking tactics that are based on a unification of the order of terms in a favorable set*, Zap. Naučn. Sem. Leningrad. Otdel. Mat. Inst. Steklov. (LOMI) **16** (1969), 126–136 = Sem. Math. V. A. Steklov Math. Inst. Leningrad **16** (1969), 64–68. MR **41** #8173.

6. ———, *The inverse method and tactics for establishing deducibility for a calculus with functional symbols*, Trudy Mat. Inst. Steklov. **121** (1972), 14–56 = Proc. Steklov Inst. Math. **121** (1972), 11–60.

7. Ju. Š. Gurevič, *Formulas with one* $\forall$, Selected Problems of Algebra and Logic, Novosibirsk, 1973, pp. 97–110. (Russian) RŽMat **1973** #11A90.

8. G. E. Minc, *Herbrand's theorem*, Math. Theory of Logical Deduction, "Nauka", Moscow, 1967, pp. 311–350. (Russian) MR **36** #4958.

9. S. Ju. Maslov, *An inverse method for establishing deducibility of nonprenex formulas of the predicate calculus*, Dokl. Akad. Nauk SSSR **172** (1967), 22–25 = Soviet Math. Dokl. **8** (1967), 16–19. MR **35** #19.

10. V. P. Orevkov, *On Glivenko sequent classes*, Trudy Mat. Inst. Steklov. **98** (1968), 131–154 = Proc. Steklov Inst. Math. **98** (1968), 147–176. MR **39** #2606; **43** #4620.

11. S. C. Kleene, *Permutability of inferences in Gentzen's calculi LK and LJ. Two papers on the predicate calculus*, Mem. Amer. Math. Soc. No. 10 (1952); Russian transl. in Mathematical Theory of Logical Inference, "Nauka", Moscow, 1967, pp. 208–236. MR **14**, 436; **36** #4957.

Translated by F. M. GOLDWARE

Trudy Mat. Inst. Steklov.
121 (1972)

Proc. Steklov Inst. Math.
121(1972)

THE SKOLEM METHOD IN INTUITIONISTIC CALCULI

UDC 51.01 : 164

G. E. MINC

ABSTRACT. The question of the admissibility of removing positive quantifiers with the help of function symbols in the intuitionistic predicate calculus is studied. Sufficient admissibility conditions are established with the help of syntactical transformations of inferences in Gentzen systems. The simplest condition formulated is: removal of a positive $\exists$-quantifier not within the scope of $\supset$ or $\neg$ is admissible. It is shown by examples that the list cited of admissible rules is maximal in a definite sense. The question of preserving consistency when removing positive quantifiers is considered separately. It is shown that the first ϵ-theorem carries over to the intuitionistic predicate calculus and the second does not. Finally, it is shown that removal of positive quantifiers, even from a prenex formula, can convert a consistent intuitionistic theory with equality into an inconsistent one. The extensionality requirement of the function symbol introduced proves to be vital here: if its fulfillment is ensured in advance (as when considering the usual method of eliminating function symbols with the help of representative predicates), removal of the quantifier is admissible. The methods developed when proving the basic results permit finding a specialization of the form of inference with the help of which the removability of structural rules in Gentzen calculi is easily proved.

Bibliography: 7 items.

CONTENTS.

Introduction

This article is devoted to the question of the admissibility in intuitionistic predicate calculus of the Skolem method of removing (eliminating) positive quantifiers with the help of function symbols. We shall consider cases when application of this method is admissible. It will be shown by examples that the list of admissible rules which we shall cite is complete (in the sense that there are no other admissible rules describable in terms of the symbols of the occurrences of quantifiers and the symbols of the occurrences of the logical connectives dominating these quantifiers). These results were announced in [1]. In §1 a specialization of the form of inference in Gentzen systems will be indicated which permits giving new proofs of the removability of the rule for

AMS (MOS) subject classifications (1970). Primary 02C15, 02G99.

cancelling repetitions. These proofs are cited in the Appendix. The basic positive result is presented in Theorem 6.1. Let us observe that the simply formulated assertion containing the basic special cases of Theorem 6.1 (Theorem 6.2) could be proved substantially more briefly by utilizing concepts connected with the "semantics" of S. Kripke. §7 contains the positive results for sequents with empty succedent, and the remaining sections are basically devoted to the negative results.

The basic method of proof in this article is applying the technique of transforming inferences in Gentzen variants of the predicate calculus and tracing the behavior of various occurrences of formulas, terms and logical symbols under such transformations. The terminology connected with this is borrowed basically from [2]. Understanding the formulations of most of our auxiliary theorems and definitions is impossible without extensive acquaintance with §§1 and 2 of that paper. In order to trace a proof it is necessary to be acquainted with the results of [2] which relate to the intuitionistic system. Terminology connected with ancestral relationships ([2], §2) will be employed especially often. The terminology employed here corresponds to the terminology of the English original of [2]. Thus the terms *"application of a rule"* and *"inference"* are used synonymously. The term *"deduction"* is employed to designate a tree in which all steps are performed by the rules of the G system, but sequents which are not axioms may be at the tops. We remind readers acquainted with [2] that an *image* of a given occurrence A in a deduction coincides with that occurrence or lies below it and is an occurrence of a formula B coinciding with A or differing from A because of replacement of terms by variables (if there are predicate inferences among the inferences *transforming* A into B); a *descendant* of a given occurrence V is an occurrence as a member of a sequent and containing some image of the occurrence V. Only the relationship "being a descendant" is defined for occurrences of logical symbols.

The differences between a formula and an image of it will play a vital role in the present paper. We introduce a number of definitions in connection with this. In a natural way we extend the relationships *"being an immediate image"* and *"being an image"* to term occurrences so that an image of an occurrence of the *characteristic term*(1) of a quantifier inference is a corresponding bound occurrence of a variable and the subterms of that term which are distinct from it do not have images.

EXAMPLE. In the proof

(1) The term t in the rules $\to\exists$ and $\forall\to$ or the variable b in the rules $\exists\to$ and $\to\forall$ (see [2], §1) is called the characteristic term of the quantifier inference.

$$
\begin{array}{ll}
E_{11}\left\{\begin{array}{r}
Pfa \to Pfa \\ \hline
\neg Pa \,\&\, Pfa \to Pfa \\ \hline
\exists x(\neg Pa \,\&\, Pfa) \to Pfa
\end{array}\right. & \\
L_1 \overline{\hspace{12em}} & \text{Distinguished system} \quad \text{Crit., } K_2 \\
\forall\beta\exists x(\neg P\beta \,\&\, Pf\beta) \to Pfa & \text{2 qu. sing.} \\
L_2 \overline{\hspace{12em}} & \text{2-sing} \\
E_0\left\{\begin{array}{r}
\neg Pfa,\ \forall\beta\exists x(\neg P\beta \,\&\, Pf\beta) \to \\ \hline
\neg Pfa \,\&\, Pffa,\ \forall\beta\exists x(\neg P\beta \,\&\, Pf\beta) \to \\ \hline
\exists x(\neg Pfa \,\&\, Pffa),\ \forall\beta\exists x(\neg P\beta \,\&\, Pf\beta) \to \\ \hline
\forall\beta\exists x(\neg P\beta \,\&\, Pf\beta),\ \forall\beta\exists x(\neg P\beta \,\&\, Pf\beta) \to \\ \hline
\forall\beta\exists x(\neg P\beta \,\&\, Pf\beta) \to
\end{array}\right. &
\begin{array}{l}
\text{2-sing.} \\ \text{2-sing.} \\ \text{Crit., } K_1 \\ \text{2-sing., 2-inference}
\end{array}
\end{array}
\quad \text{2-singular pairs}
$$

(the auxiliary indicators will be explained below) the occurrence of the term $f\beta$ in the bottom sequent is an image of the left occurrence of the term fa in the top sequent, as well as an image of the unique occurrence of the term ffa. Every formula occurrence has an image (moreover, a unique one) in any lower-lying sequent. At the same time a term occurrence V in a premise of a predicate inference (for example, the occurrence of the variable a in the third sequent from the bottom) may not have an image in the conclusion. This can happen if a minus-inference(2) is being considered and V is contained strictly inside of the term occurrence on which the quantifier is "suspended".

LEMMA 0.1. *Let V be an occurrence of a term r in a deduction D, and U the lowest image of V in D. Then U is either an occurrence in the bottom sequent of the deduction D or is an occurrence in a premise of a minus-inference whose characteristic term contains r and is different from r.*

The lemma is proved by induction on the height of the occurrence V.

When considering break-downs of sequents into parts, the notation $S_1 \cdot S_2 \cdot \cdots \cdot S_n$ will be employed for a sequent whose antecedent is the union of the antecedents of the sequents $S_1, \cdots, S_n$ and whose succedent is the union of the succedents of those sequents. We shall denote by T_a the sequent obtained from T as the result of cancelling its succedent. We shall say that V is an occurrence as the *principal symbol* of the formula A if V is an occurrence as the principal logical connective of the formula A or if A is an elementary formula and V is the (unique) predicate symbol occurrence in A.

In the sequel the need will often arise for encompassing one-premise and two-premise rules in a common schema. In such cases notation of the form

$$\frac{S_1\,(S_2)}{S_3} \tag{1}$$

(2) We call inferences by the rules $\to\forall$ and $\exists\to$ (the rules for introducing positive quantifiers) plus-inferences, and inferences by the rules $\forall\to$ and $\to\exists$ minus-inferences.

will be used or, more briefly,(3)

$$S_1(S_2) \vdash S_3.$$

In case the rule under consideration is two-premise, (1) denotes the figure

$$\frac{S_1\,S_2}{S_3},$$

and in case the rule under consideration is one-premise, the figure

$$\frac{S_1}{S_3}.$$

By the *frame* of any rule of any of the calculi under consideration we mean the rule obtained after cancelling all parametric formula occurrences. Any rule of the calculus G, except the rules $\neg\rightarrow$ and $\supset\rightarrow$ of the intuitionistic system, can (to within permutations) be written as

$$U \cdot W_1,\ (U \cdot W_2) \vdash U \cdot P, \tag{2}$$

where

$$W_1,\ (W_2) \vdash P \tag{3}$$

is the frame of the rule under consideration. Any rule of G can (to within permutations) be written in the form $U^0 W_1(U \cdot W_2) \vdash P$, where (3) is the frame of the rule under consideration and U^0 denotes U_a if the rule under consideration is one of the rules $\neg\rightarrow$ or $\supset\rightarrow$ of the intuitionistic system, and otherwise denotes U.

§1. Degenerate applications of rules and specialization of the form of proof

Let us say that the formula A is obtained from the formula B by *trivial cancellation* if B has the form $Q_1x_1 \cdots Q_nx_nA$, where $n \geqslant 0$, and $Q_1x_1, \cdots, Q_nx_n$ are degenerate quantifiers, i.e. $x_1, \cdots, x_n$ do not occur free in A.

Let D be a deduction in the calculus G, and L a logical inference in D. Let us call L *semidegenerate* if L does not introduce a degenerate quantifier and at least one of the following conditions is fulfilled:

a) $L \neq (\exists\rightarrow)$ and at least one of the side antecedent formulas of L is obtained by trivial cancellation from an antecedent member, different from the principal formula of L, of some sequent below a premise of L.

b) $L = (\exists\rightarrow)$ and below a premise of L there is an antecedent member, different from the principal formula of L, from which a formula of the form(4) $[A]^x_r$ is obtained by trivial cancellation, where $\exists xA$ is the principal formula of L and r is a term free for x in A.

We call a semidegenerate inference different from $\rightarrow\supset$ *degenerate.*

---

(3) This usage of the symbol $\vdash$ differs somewhat from its usual meaning of "is provable".

(4) By $[A]^{x_1\ldots x_n}_{t_1\ldots t_n}$ is denoted the result of successive substitution into A of the term t_1 for all free occurrences of $x_1, \cdots,$ of the term t_n for all free occurrences of x_n.

We shall prove in this section that it is possible to remove degenerate inferences from a deduction.

The operation of *multiplying a sequent* by a proof or deduction in the system G will play an important role in the sequel. We shall denote by $T \cdot D$ the result of applying this operation to the sequent T and the deduction D. If the succedent of T is not empty or the characteristic variable of any plus-inference of D occurs in T, by definition $T \cdot D$ coincides with D. If, however, this condition is not fulfilled, the figure $T \cdot D$ is obtained from D by multiplying each sequent of D by T.

Reviewing the rules of G, we ascertain that multiplication by a sequent converts each application of a rule into an application of the same rule, with the same principal term and side formulas.

Let us call a deduction D in G *nondegenerate* if there are no degenerate inferences in D and an antecedent side formula of each semidegenerate $(\to \supset)$-inference is introduced by thinning.

THEOREM 1.1. *Let D be a pure deduction of the sequent S from the sequents $S_1, \cdots, S_k$ in the G system. Then D can be reconstructed into a nondegenerate deduction D' such that for some lists of variables $X_1, \cdots, X_k$ and lists of terms $Y_1, \cdots, Y_k$, D' is a deduction of S from sequents differing from $[S_1]^{X_1}_{Y_1}, \cdots, [S_k]^{X_k}_{Y_k}$ by at most only trivial cancellations.*

Let us prove the theorem by induction on the number of logical inferences in the given deduction. The basis is obvious. Let us justify the induction step by induction on the number b of semidegenerate $(\supset \to)$-inferences an antecedent side formula of which is not introduced directly by thinning.

Case 1. There is a degenerate inference in the deduction. Let us find one such inference and denote it by L. Let us find the lowest sequent occurrence in D which jointly with L satisfies the condition appearing in the definition of a semidegenerate inference and denote it by V. Let us denote by E that part of the deduction which terminates with the occurrence V. The deduction E has (to within permutations of members of sequents) the following form:

$$
\begin{array}{ll}
 & \quad \Downarrow \qquad\quad \Downarrow \\
D_2 & \underline{W_1U^0 \qquad (W_2U)}\,L \\
 & \qquad\quad PU \\
D_1 \left\{ \begin{array}{l} \\ \\ \end{array} \right. & \qquad\quad \Downarrow \\
 & \qquad\quad W'_\delta R
\end{array}
\qquad\qquad \delta \leqslant 1.
$$

The notation introduced here is intelligible from the schema: W'_δ is a sequent consisting of a unique antecedent member; to within trivial cancellation it coincides with W_1 (then $\delta = 1$), with W_2 $(\delta = 2)$ or (if L is $\exists \to$) is obtained from W_1

by substituting some term t for all free occurrences of x (where x is such that the principal formula of L is $\exists xA$). For simplicity of notation, let us assume that $\delta = 1$ and trivial cancellations are not needed. Let us replace the deduction E with the following figure:

$$\left.\begin{array}{c}(\text{У})\,\dfrac{W_1'U^0}{W_1'PU}\end{array}\right\}D_2^*$$

$$\left.\begin{array}{c}\Downarrow\\[4pt](\text{C})\,\dfrac{W_1'W_1'R}{W_1'R}\end{array}\right\}W_1D_1$$

Here D_2^* coincides with D_2 if W_1' coincides with W_1; if W_1' is $[W_1]_t^x$, it is obtained from D_2 by substituting the term t for all occurrences of the variable b in D_2, where b is the characteristic variable of L. (If trivial cancellations are needed, all explicitly written out occurrences of W_1' in the new figure, except the lowest occurrence, must be replaced by the result of the corresponding trivial cancellation, and the necessary rule applications introducing degenerate quantifiers must be inserted.) By virtue of deduction D's purity, the formula W_1' is, as a matter of fact, in the antecedent sequents of D_1. D_2^* is a deduction, again by virtue of deduction D's purity.

Thus the whole figure referred to is a deduction, L being replaced in it by structural inferences.

Case 2. There are no degenerate inferences in the deduction. If $b = 0$ (there are no semidegenerate inferences), everything is proved. If $b > 0$, the required transformation can be accomplished in accordance with the schema

Original figure

$$\begin{array}{rc}D_2\left\{\begin{array}{c}\Downarrow\\ (A \to B)\,U\end{array}\right.\\ \hline D_1\left\{\begin{array}{c}\to (A \supset B)\,U\\ \Downarrow\\ (A \to\ )\,R\end{array}\right.\end{array}$$

Altered figure

$$\begin{array}{rc}D_2\left\{\begin{array}{c}\Downarrow\\ (A \to B)\,U\end{array}\right.\\ (A \to\ )\cdot D_1\left\{\begin{array}{c}\dfrac{(AA \to B)\,U}{(A \to A \supset B)\,U}\\ \Downarrow\\ \dfrac{(AA \to\ )\,R}{(A \to\ )\,R}\end{array}\right.\end{array}$$

The theorem is proved.

We need further specialization of the form of inference. In connection with this let us recall some definitions. A proof is called *weeded* (cf. [3]) if it possesses the property of Lemma 4 of [2], i.e. if all thinnings are dropped maximally downwards. In a weeded proof each occurrence of a formula as a member of a sequent is traceable to axioms or is introduced directly by thinning. Reviewing the proof of Lemma 4 of

[2], we see that *with respect to every proof it is possible, by cancelling certain sequents and certain members of sequents and adding structural inferences, to construct a weeded proof with the same endsequent.*

Let us say that D satisfies Condition (A) if D is a pure weeded nondegenerate proof, the formulas occurring in the axioms of D are elementary, and the sequent members which begin with degenerate quantifiers are introduced directly by thinning or predicate inference. Combining what is said above with Theorem 1.1, we obtain the following assertion.

THEOREM 1.2. *It is possible to construct with respect to every pure proof a proof with the same endsequent which satisfies Condition* (A). *If in this connection the original proof is nondegenerate and the formulas occurring in the axioms are elementary, then in the process of reconstructing it is possible to get along with transformations of the following types*:

1) *cancellation of certain sequents and certain members of sequents*;
2) *insertion of structural inferences*;
3) *insertion of predicate inferences introducing degenerate quantifiers.*

PROOF. By virtue of Lemma 9 of [2] it can be assumed that formulas occurring in axioms are elementary. By virtue of Theorem 1.1 it can be assumed that the proof is nondegenerate. We achieve fulfillment of the condition on predicate inferences by cancelling from the given proof all outer occurrences of degenerate quantifiers (i.e. the occurrences with which formulas, as members of sequents, begin), except for the lowest such occurrences. The figure obtained becomes a proof after insertion of the corresponding predicate inferences. Then, applying Lemma 4 of [2], we perform a "weeding". As the result, a proof is obtained which satisfies all items of Condition (A), except, perhaps, for the condition on predicate inferences. It can be violated because of the insertion of structural inferences. By permuting the "nuisance" structural inferences with predicate inferences interesting us we obtain a proof satisfying Condition (A).

§2. The Skolem method of eliminating quantifiers

Let us call a sequent *purified* if no variable occurs both free and bound in it, there are no degenerate quantifiers, and different occurrences of quantifiers contain distinct variables. Let us call a variable x occurring bound in a purified sequent S *S-positive* if x occurs in a positive(5) quantifier, and *S-negative* otherwise.

Let S be a purified sequent and x an S-positive variable. The following condition on x and S will play an essential role in the sequel:

(5) An occurrence of the $\forall$-quantifier ($\exists$-quantifier) is taken as *positive* if it is within the scope of an even (respectively, odd) number of occurrences of negation and premises of implications. An occurrence is *negative* if it is not positive. In view of the fact that purified sequents are under consideration, it is possible to speak of the signs of quantifiers and not of the signs of their occurrences.

CONDITION (+). *If some S-positive* $\forall$*-quantifier dominates the quantifier* Qx *or coincides with it, this* $\forall$*-quantifier is essentially positive.*(6)

This condition is satisfied in the first place by essentially positive occurrences of $\forall$-quantifiers, and secondly by those negative occurrences of $\exists$-quantifiers which positive $\forall$-quantifiers do not dominate. We shall usually remove quantifiers by beginning with the outermost. In connection with this let us introduce

CONDITION (++). (+) *holds, and no S-positive quantifier dominates the quantifier* Qx.

It will be assumed in the sequel (up to §6) that S and x satisfy Condition (++).

Let us write the complete list (without repetitions) of characteristic variables of the S-negative quantifiers dominating the quantifier Qx in the form $\beta_1, \cdots, \beta_k$, $k \geqslant 0$. Let us write the scope of Qx as QxA. Let f be the first, in alphabetical order, k-place function symbol not occurring in A. Let us call the sequent obtained from S after replacing QxA by $Qx[A]^x_{f(\beta_1,\dots,\beta_k)}$, the result of applying the Skolem method of eliminating quantifiers (abbreviated as SME) to the pair (S, x), and denote it by $\mathfrak{S}(S, x)$.

EXAMPLE 1. If S is $\forall\beta\exists x(\neg P\beta \mathbin{\&} Px)\rightarrow$, then $\mathfrak{S}(S, x)$ is $\forall\beta\exists x(\neg P\beta \mathbin{\&} Pf(\beta))\rightarrow$.

Let us note that such a definition of the operation $\mathfrak{S}$ differs from the one introduced in [1] in that $\mathfrak{S}(S, x)$ *contains a degenerate quantifier with respect to* x. This difference is evoked by technical considerations.

By the theorem on monotone replacement in [4], the provability of S implies the provability of $\mathfrak{S}(S, x)$. In the case of the classical predicate calculus the provability of $\mathfrak{S}(S, x)$ implies the provability of S. We shall prove below that under certain additional conditions the latter also holds for the intuitionistic case.

Let us introduce the notation

$$t_{1\dotplus k} \leftrightarrows t_1, \ldots, t_k$$

$$A(t_1, \ldots, t_k) \leftrightarrows [A]^{\beta_{1\dotplus k}}_{t_{1\dotplus k}}$$

$$A(t_1, \ldots, t_k, t_{k+1}) \leftrightarrows [A]^{\beta_{1\dotplus k}x}_{t_{1\dotplus k}t_{k+1}}$$

Let D be an arbitrary proof of the sequent $\mathfrak{S}(S, x)$. Let us call an inference in D *critical* if one of the side occurrences of this inference begins with Qx and is traceable to axioms. (Let us call this side occurrence an f-occurrence.) We call the premise of a critical inference which contains the f-occurrence the *basic premise.* If the basic premise is the conclusion of the predicate inference by which Qx is introduced (the f-occurrence is obtained), let us call that predicate inference *hypercritical,* and let us call a figure consisting of critical and hypercritical inferences a *critical figure.*

---

(6) An $\forall$-quantifier is essentially positive if it is neither within the scope of a negation nor in the premise of an implication.

Let us observe that by virtue of the hypothesis that no positive quantifier dominates the quantifier Qx, any formula of D which begins with Qx (in particular, the base(7) of any f-occurrence) has the form $QxA(t_{1\div k}, f(t_{1\div k}))$. We call the term $f(t_{1\div k})$ *the term belonging to the critical inference* under consideration. Let us observe that if D satisifes Condition (A) of §1, the basic premise of any critical inference is actually the conclusion of the quantifier inference introducing Qx, since an f-occurrence is traceable to axioms and begins with a degenerate quantifier.

EXAMPLE 2. The proof of the sequence $\mathcal{S}(S, x)$ which is cited in the Introduction, where S is the sequent of Example 1, contains two critical inferences (both $(\forall\rightarrow)$-inferences). The term fa belongs to the upper critical inference, and the term ffa, to the lower one.

The basic step in reconstructing a proof D of a sequent $\mathcal{S}(S, x)$ into a proof of the sequent S consists in carrying out the transformation described below.

Let us write down all terms belonging to critical inferences in the list

$$r_1, \cdots, r_m, \tag{1}$$

in which longer terms precede shorter ones. Let us fix those m of the mutually distinct variables $a_1, \cdots, a_m$ which do not occur in D. If E is a term, a formula, a list of formulas, or a sequent, E^* will denote $[E]^{r_1 \ldots r_m}_{a_1 \ldots a_m}$. Let us replace each occurrence in D of a formula of the form $QxA(t_{1\div k}, f(t_{1\div k}))$ by the formula $QxA(t_{1\div k}, x)$ and after that replace each sequent S by S^*; i. e. let us replace all occurrences of r_1 by $a_1, \cdots$, and all occurrences of r_m by a_m. Let us denote the described operation by R and the result of applying it to the proof D, by $R(D)$.

EXAMPLE 2 (continuation). The list (1) has the form $\{ffa, fa\}$, and $R(D)$ is

$$\begin{array}{r}
Pa_2 \rightarrow Pa_2 \\ \hline
\neg Pa \& Pa_2 \rightarrow Pa_2 \\ \hline
\exists x(\neg Pa \& Px) \rightarrow Pa_2 \\ \hline
\forall\beta\exists x(\neg P\beta \& Px) \rightarrow Pa_2 \\ \hline
\neg Pa_2, \forall\beta\exists x(\neg P\beta \& Px) \rightarrow \\ \hline
\neg Pa_2 \& Pa_1, \forall\beta\exists x(\neg P\beta \& Px) \rightarrow \\ \hline
\exists x(\neg Pa_2 \& Px), \forall\beta\exists x(\neg P\beta \& Px) \rightarrow \\ \hline
\forall\beta\exists x(\neg P\beta \& Px), \forall\beta\exists x(\neg P\beta \& Px) \rightarrow \\ \hline
\forall\beta\exists x(\neg P\beta \& Px) \rightarrow
\end{array} . \tag{2}$$

Let us see how operation R introduces changes into critical figures. Each such figure can be written in the form

$$\begin{array}{l}
M^0 \cdot W \\ \hline
M^0 \cdot W_1 \; (M \cdot W_2) \\ \hline
\quad MP
\end{array} , \tag{3}$$

(7) The base of an occurrence V is the word an occurrence of which is V.

where $W_1(W_2) \vdash P$ is the frame of the critical inference. The unique formula which the sequent W_1 contains as a member has the form $QxA(u_{1\div k}, f(u_{1\div k}))$, where $u_1, \cdots, u_k$ are terms. After we carry out transformation R, figure (3) turns into the figure

$$\frac{\dfrac{M_1^{0*}W^*}{M_1^{0*}W_1^*}\,\pi_1 \quad (M_1^*W_2^*)}{M_1^*P_1^*}\,\pi, \tag{4}$$

where the sequent M_1 is obtained from M by substituting formulas of the form $QxA(t_{1\div k}, x)$ for occurrences of formulas of the form $QxA(t_{1\div k}, f(t_{1\div k}))$. The sequent P_1 is obtained from P by substituting the formula $QxA(u_{1\div k}, x)$ for the occurrence of the formula $QxA(u_{1\div k}, f(u_{1\div k}))$ as the principal logical component of the only formula occurring in P. Finally, the sequent W_1^* (the sequent W^*) consists of a single member, the formula $QxA(t_{1\div k}^*, x)$ (respectively $A(t_{1\div k}^*, a_i)$). Let us denote by π the figure which a critical inference turns into, and by π_1 the figure a hypercritical inference turns into. π is an inference by the same rule as the critical inference. π_1 is an $(\rightarrow\forall)$- or $(\exists\rightarrow)$-inference only if the restriction on variables is fulfilled, i.e. if π_1 does not occur in $M_1^{0*}W_1^*$. Let us call the critical inferences for which this condition is fulfilled *regular*.

EXAMPLE 2 (continuation). The lower critical inference in the proof in the Introduction is regular. The $(\exists\rightarrow)$-inference in (2) serves as the corresponding hypercritical inference. The upper critical inference is irregular. The restriction on variables is violated in the corresponding position of (2).

Below (in §5) syntactical conditions will be indicated which are sufficient to make it possible to construct with respect to every proof of a sequent $\mathfrak{S}(S, x)$ a proof of it in which all critical inferences are regular.

Let us consider how various inferences are altered as the result of transformation R. Replacing $QxA(t_{1\div k}, t_{k+1})$ by $QxA(t_{1\div k}, x)$ spoils only inferences with principal formula $QxA(t_{1\div k}, t_{k+1})$, i.e. hypercritical inferences. Moreover, as already established, hypercritical inferences corresponding to regular critical ones turn into predicate inferences.

Let us prove that successive replacement of $r_1, \cdots, r_m$ by $a_1, \cdots, a_m$ does not spoil any inferences. To this end let us ascertain that the conditions of Lemma 2 of §4 of [5] are fulfilled for minus-inferences. That lemma is formulated for $LK^=$, i.e. for a calculus with two kinds of variables (free and bound) in which multimember succedents are admitted. However, the first of these differences from our calculus is absorbed by the requirement for purity of variables (as a result of which we can classify the variables occurring in a given inference into free and bound), and the second, by the fact that replacing a term by a term does not alter the number of members of a succedent.

Hence let us assume that when replacing a term r of the list (2) by a variable a for some inference π (with characteristic term t and characteristic variable y) the condition of the lemma is violated, i.e. a term r' containing the variable y, differing from the latter variable and being such that r coincides with $[r']^y_t$, occurs in the principal formula of π.

Since r begins with f and r' is different from y, r' begins with f. Since r' begins with f and contains the bound variable y, all occurrences of r' must be within the scope of the quantifier Qx. (This property holds for the lower sequent and "is inherited" during the ascent.) However, in this case r' (or a term containing it) must be replaced by x in the first step of transformation R; namely, when replacing $QxA(u_{1\div k}, f(u_{1\div k}))$ by $QxA(u_{1\div k}, x)$.

We have thus ascertained that transformation R spoils perhaps only irregular hypercritical inferences, and by the same token have proved the following assertion.

THEOREM 2.1. *If Condition* $(++)$ *is fulfilled, then with respect to every proof of a sequent* $\mathfrak{S}(S, x)$ *it is possible to construct a proof of the sequent* S.

Methods for "regularizing" a proof will be considered in the following section.

§3. Irregularity condition for critical figures

The terminology and stipulations of §2 are applied in this section.

LEMMA 3.1. *Let Condition* $(++)$ *be fulfilled, and let* D *be a proof of the sequent* $\mathfrak{S}(S, x)$. *Let* π *and* π' *be two inferences in* D, *images of the principal formulas of* π *and* π' *coinciding in a lower sequent of* D, *and let there be critical inferences* π_k *and* π'_k *to which the same term belongs and of whose principal formulas the principal formulas of respectively* π *and* π' *are descendants. Then the principal (and consequently also the side) formulas of* π *and* π' *coincide.*

Indeed, any occurrence of a formula differs from its image perhaps only by the replacement by certain terms of the variables bound by the quantifiers governing this image but not occurring in it. By virtue of Condition $(++)$, these variables for the formulas under consideration are contained in the list $\beta_1, \cdots, \beta_k$, and therefore all replacements of these variables by the terms $u_1, \cdots, u_k$ are "recorded" in the term $f(u_1, \cdots, u_k)$ which belongs to the corresponding critical inference.

LEMMA 3.2. *Let* S *and* x *satisfy Condition* $(++)$, *let* D *be a proof of the sequent* $\mathfrak{S}(S, x)$ *which satisfies Condition* (A), *and let* π *be an irregular critical inference in* D. *Then below* π *there is a minus-inference whose characteristic term contains the term belonging to* π.

PROOF. Let us denote by r the term belonging to π. Let us observe that r does not contain the variables $\beta_1, \cdots, \beta_k$. (They are all bound, and D is a pure proof.) The term r does not occur in the bottom sequent of D, since the only term beginning with f and occurring in that sequent contains $\beta_1, \cdots, \beta_k$. Let us assume that there is no minus-inference below π whose characteristic term contains r. Let us prove that in this case Q is $\exists$ and there is a critical inference π_1 below π

such that the terms belonging to π and π_1 coincide. The proof of our lemma will be completed after this: π turns out to be a semidegenerate inference (since Q is $\exists$, π has an antecedent side formula; by virtue of Lemma 3.1, the inference π_1 has the same side formula), which contradicts Condition (A). Since only the term occurrences which "vanish" under applications of minus-rules may not have images, each occurrence of r in the basic premise of the inference π has, in the case under consideration, an image in the bottom sequent. The base(8) of this image is $f(\beta_{1\div k})$. All occurrences of f in the bottom sequent are contained within the scope of Qx. (Let us denote it by U_x.) This implies that each occurrence of the term r in the basic premise and beneath it is contained in some occurrence of a formula whose image in the bottom sequent of the proof D intersects with U_x. All occurrences of r in the basic premise which are within the scope of Qx are replaced by the variable x when realizing the operation R. Since π is not regular, there is a parametric occurrence H of a formula as a member of the basic premise of π which is not within the scope of Qx and contains r. Let us denote by H' the image of H in the bottom sequent. As already noted, H' intersects U_x. Since H does not contain Qx, H' does not contain U_x. Therefore H' is contained in U_x. Hence below π there is an inference π_1 one of whose side occurrences is a preimage of an occurrence of U_x, with H belonging to this side occurrence. This now implies that Q is $\exists$. Indeed, if this were not so, then by virtue of Condition (++) the occurrence of Qx in S (and hence also the occurrence of U_x) would be essentially positive, which is impossible in view of the presence of the pair of inferences π and π_1 located in the same branch and side occurrences of which are ancestors of the occurrence of U_x.

π_1 is a critical inference, since otherwise a side occurrence of it which is a preimage of U_x would not be traceable to axioms and (by virtue of the weededness of D) would not have a preimage in a premise of π.

Finally, r (the term belonging to π) coincides with the term t belonging to π_1 because the base of the occurrence H contains r, while the base of the f-occurrence of the inference π_1 contains an image of H which contains r. (Let us recall that, by hypothesis, below π (a fortiori below π_1) there are no minus-inferences whose characteristic terms contain r; therefore r cannot be the characteristic subterm of t.)

§4. Diminishing the irregularity of proofs

According to Theorem 2.1, to construct a proof of a sequent S it is sufficient to have a proof of $\mathcal{S}(S, x)$ in which all critical inferences are regular. Having an irregular proof, we attempt to diminish its "irregularity". To this end we must introduce some measure of irregularity. By virtue of Lemma 3.2 it can be assumed that a critical inference π is irregular only if there is a minus-inference below it whose

(8) The word of which V is an occurrence is called the base of V.

characteristic term contains the term belonging to π. To remove irregularity it would be sufficient to move these "bad" minus-inferences higher than the respective critical figures. The inferences separating the critical figures and the "bad" minus-inferences hinder such permutation. A measure of the irregularity of a proof will be the sum (with definite weights) of the numbers of permutations necessary for the regularization of all critical figures.

Let us pass to precise definitions. As in §2, we write in the form $r_1, \cdots, r_m$ the list of all terms belonging to the critical inferences of D (in which longer terms precede shorter ones). Let us call a minus-inference an *i-inference* $(i = 1, \cdots, m)$ if its characteristic term contains r_i. Thus the term r_i appears just after the i-inferences in our proof (if it is looked at from below). Let us say that a logical inference π belongs to the *class* K_i if there is a critical inference to which the term r_i belongs and the principal occurrence of which belongs to the principal occurrence of π. (Inferences of the class K_i "break up" large formulas until the principal formula of a critical inference is obtained in which r_i appears.) Let us call a *pair* of logicalities π_1/π_2 *i-singular* if π_1 is located higher than π_2, $\pi_1 \in K_i$, $\pi_2 \notin K_i$ and there is an i-inference located below π_2 or coinciding with π_2. Let us say that an *inference* is *i-singular* if it is the lower member of an i-singular pair (exactly the singular inferences hinder regularization), and that an *inference* π is *i-quasisingular* if $\pi \in K_i$ and there is an i-inference below π. (In the process of reconstructing a proof i-quasisingular inferences will not turn into i-singular ones.) Let us call an object of the type under consideration (a pair of inferences, an inference), singular (quasisingular) if it is i-singular (respectively i-quasisingular) for some i. The concepts we have defined are illustrated by the example in the Introduction.

For each $i = 1, \cdots, n$ let us denote by $l_i(D)$ the number of distinct i-singular pairs. Let us call the "transfinite number" $\Sigma_1^n l_i(D)\,\omega^{n-i}$ the *height* of the proof D. The height of a proof will also serve as a measure of its "irregularity".

The figure below explains the collocation of the members of singular pairs. Inferences which can, generally speaking, coincide with each other are marked with the same number of horizontal strokes.

$=$ critical inference to which
$\downarrow$ the term r_i belongs
$=$ inference of class K_i
$\downarrow$
$=$ i-quasisingular inference — i-singular pair (1)
$\downarrow$
$-$ i-singular inference
$\downarrow$
$-$ i-inference $(\overline{\forall \to}, \to \exists)$

Let us explain the arrangement of the marks: an i-singular inference does not belong to K_i and therefore can be neither i-quasisingular nor critical; an i-inference cannot belong to the class K_i; the characteristic term of any critical inference κ has the form $[f(\beta_1, \cdots, \beta_k)]^{\beta_1 \ldots \beta_k}_{t_1 \ldots t_k}$, where $\beta_1, \cdots, \beta_k$ are minus-variables and $t_1, \cdots, t_k$ are the terms substituted when throwing out these variables. If the principal formula of the inference κ belongs to the principal occurrence of some minus-inference π, then π consists of throwing out a quantifier with respect to β_j with the substitution of the term t_j (for some j). Therefore, if an i-inference were to belong to the class K_i, the term r_i (the characteristic term of the corresponding critical inference) would contain itself as the characteristic part (namely, as the part of some t_j).

The possibility of realizing the permutation we need will be proved with the help of the permutability theorems in [2]. As was noted there, permutation is not always possible in a proof in the intuitionistic system G. In particular, the pairs

$$\frac{(\supset\to),\ (\neg\to)}{(\to\supset),\ (\to\neg)} \qquad \frac{(\supset\to),\ (\neg\to),\ (\to\vee),\ (\to\exists)}{\vee\to} \tag{2}$$

$$\frac{(\to\supset),\ (\to\&),\ (\to\neg)\ (\to\forall)}{\vee\to} \tag{3}$$

$$\frac{(\forall\to),\ (\to\exists)}{(\exists\to)} \qquad \frac{(\forall\to)}{(\to\forall)} \tag{4}$$

appear in Lemma 7 of [2] as exceptional.

Sufficient permutability conditions are given in [2] for exceptional pairs of the forms (3) and (4) (cf. Lemma 8 and footnote 15); these will be fulfilled in the conditions of Theorem 4.1 cited below. Let us call pairs corresponding to (2), i. e.

$$(\supset\to)/(\to\supset),\quad (\supset\to)/(\to\neg),\quad (\neg\to)/(\to\supset),\quad (\neg\to)/(\to\neg),$$
$$(\supset\to)/(\vee\to),\quad (\neg\to)/(\vee\to),\quad (\to\vee)/(\vee\to),\quad (\to\exists)/(\vee\to),$$

exceptional.

We shall say that a type(9) σ is *x-exceptional* (relative to S) if there is an exceptional pair τ/σ such that some occurrence of a logical symbol belonging to type τ dominates the occurrence of Qx in S.

THEOREM 4.1. *Let a sequent S and an S-positive variable x satisfy Condition* (++). *Let a proof D of the sequent $\mathfrak{S}(S, x)$ satisfy Condition* (A), *and let neither singular nor quasisingular inferences belonging to an x-exceptional type be in D. Then by permuting rule applications and applying the transformations mentioned in the*

(9) In other words, a symbol of the form $(\to\alpha)$ or $(\alpha\to)$, where α is one of the symbols $\&$, $\vee$, $\supset$, $\neg$, $\forall$ or $\exists$. Positive occurrences of a symbol α are called *occurrences of type* $(\to\alpha)$, and negative ones, *occurrences of type* $(\alpha\to)$.

formulation of Theorem 1.2 *it is possible to construct a proof of the same sequent which satisfies Condition* (A) *and does not contain singular inferences.*

We prove the theorem by induction on the height of the proof under consideration. Let us make a few remarks.

REMARK 1. The usual permutations of rule applications do not change the list of terms belonging to critical inferences. As to permutations with redesignation of the characteristic variables of plus-inferences (for example, permutations of $\vee\rightarrow$ with a pair of $\rightarrow\forall$ or $\exists\rightarrow$ above it), all occurrences of terms subject to alteration change simultaneously (by virtue of the purity of variables) under the corresponding redesignation so that either the list interesting us turns into an isomorphic one or some members are "identified" in it. In any case it is not lengthened, and its members do not change places. The remaining transformations mentioned in the formulation of the theorem are just as "conservative".

REMARK 2. By virtue of the theorem's condition pertaining to types of singular inferences, any singular pair in D is different from any of the pairs (2).

REMARK 3. Permutations of inferences do not introduce degenerate inferences and do not violate the properties of purity of variables and weededness. The only part of Condition (A) which may turn out to be violated, namely the restriction on predicate inferences introducing degenerate quantifiers, is restored in accordance with Theorem 1.2.

Let us pass to the proof of the theorem. The basis of the induction is trivial. The proof of the induction step will be carried out by analyzing several cases. In each of these cases the basic role will be played by some pair or triple of adjacent logical inferences which we shall call the *distinguished system.* In case the distinguished system turns out to be a triple, its bottom inference will be two-premise, with one of its remaining inferences located above the left premise and the other above the right. Let us denote by L_1 the top (left top if the distinguished system is a triple) inference belonging to the distinguished system, and by L_2 the bottom inference. If the distinguished system is a triple, let us denote by (L_1) the right top inference belonging to that system. Let us define some break-down of the proof D into parts (cf. the schema cited below). Let us denote by E_0 the figure obtained if all sequents above the conclusion of L_2 are removed from D. Let us denote by E_{11} (by E_{12}) the figure consisting of the left (respectively, the right) premise of L_1 and all sequents above it. In case the distinguished system is a triple and L_1 is a two-premise inference, the proof D has the form

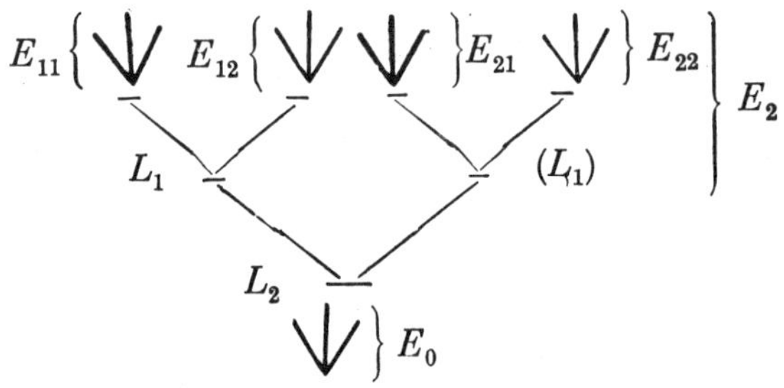

(5)

The meaning of the notation E_2, E_{21} and E_{22} is clear from the schema.

When justifying the induction step the basic device will be permuting the bottom member of the distinguished system with its top member (in case the distinguished system is a pair) or with both top members simultaneously (in case the distinguished system is a triple). As the result of such permutation the figure cited above takes the form

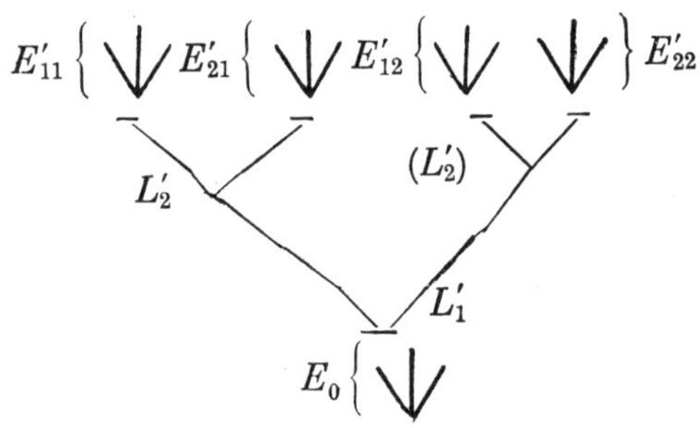

(6)

The figures $E'_{11}, E'_{21}, E'_{12}$ and E'_{22} differ respectively from E_{11}, E_{21}, E_{12} and E_{22} only by possible insertions or cancellations of structural inferences. L'_1 is an inference by the same rule as L_1, with the same principal and side formulas; L'_2 and (L'_2) are inferences by the same rule as L_2, again with the same principal and side formulas. In other cases (if, for example, the distinguished system is a pair and not a triple, or L_1 is a one-premise inference) the schemata of the result of permuting are obtained by obvious alterations of schema (6); they will be cited when considering the respective cases.

In the case when several parts $F_1, \cdots, F_n$ are distinguished in the proof we shall classify pairs of inferences, one of which is placed higher than the other, by the membership of the pair's members in parts of the break-down. The assertion that *a given pair belongs to type* F_i/F_j $(i, j = 1, \cdots, n)$ means that the upper member of the pair belongs to F_i and the lower belongs to F_j.

The distinguished system will be chosen so that the pair L_1/L_2 (and the pair $(L_1)/L_2$ in the case of the presence of (L_1)) will be i-singular $(1 \leqslant i \leqslant m)$.

Therefore $L_1, (L_1) \in K_i$, while $L_2 \notin K_i$, so that the images of the principal occurrences of L_1 and (L_1) in the conclusion of L_2 do not intersect with the principal occurrence of L_2.

Let us denote by D_1 the proof obtained from D as the result of the above permutation. The correspondence between parts of D and of D_1 which was established when describing the permutation induces a correspondence between the logical inferences in D and the logical inferences in D_1 such that the principal and side formulas of the inferences corresponding to each other coincide.

To justify the induction step, it is sufficient to prove that if the height of D is different from zero, it is possible to choose the distinguished system and to realize the permutation in such a way that the height of D_1 will be less than the height of D and the restriction on types of singular inferences will be met. As a matter of fact, it is possible in this case, by virtue of Remark 3, to obtain a proof D_2 satisfying Condition (A). The fact that the transformations mentioned in the formulation of our theorem preserve both the collocation of the remaining logical inferences and their principal and side formulas implies that the height of D_2 is not greater than the height of D_1.

Let us observe that the restriction on x-exceptional types is met for D_1 (hence also for D_2). It could be violated in the new proof only on account of L_1', L_2' or (L_2') since the remaining singular (quasisingular) inferences of D_1 have singular (quasisingular) preimages in D. But L_1' has the quasisingular preimage L_1 in D, and L_2', the singular preimage L_2. Therefore, if our restriction were violated in D_1, it would be violated in D also.

Let us justify the induction step. Let us denote by p the least i $(1 \leqslant i \leqslant m)$ such that $l_i > 0$. There are p-singular pairs in D and even p-singular pairs of adjacent inferences. (The latter can be found by gradually diminishing the distance between the members of an arbitrary p-singular pair.)

Case 1. There is a p-singular pair of adjacent inferences in D whose bottom member is one-premise. Let us declare any such pair the distinguished system. Since L_2 is one-premise (the notation of schema (5) is being employed), L_1/L_2 does not have any of the forms (3). Let L_1/L_2 have one of the forms (4). By virtue of footnote 15 of [2], it is sufficient in this case to prove that the characteristic term of the inference L_1 (we denote it by t) does not contain the characteristic variable of the inference L_2 (we denote it by a). Since the pair under consideration is p-singular, there is a p-inference, i.e. an $(\forall \rightarrow)$- or $(\rightarrow \exists)$-inference, below L_2 whose characteristic term contains r_p. Since $L_1 \in K_p$, r_p contains t. If t were to contain the variable a, the same would be true for r_p, and therefore the proof D would not be pure. Hence permuting L_1 and L_2 is always possible in the case under consideration. Let us do it:

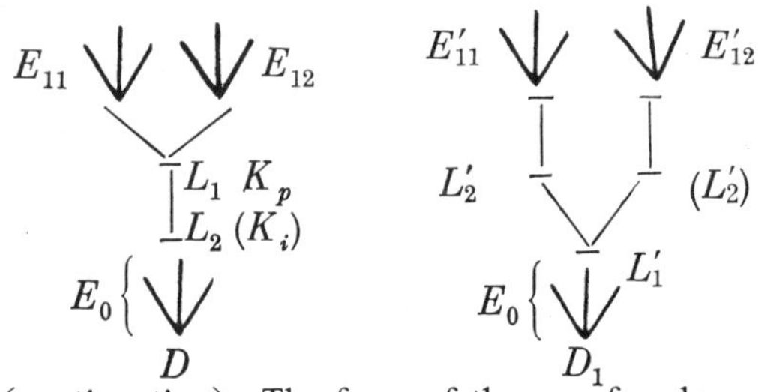

EXAMPLE 2 (continuation). The form of the proof under consideration is

After permutation

$$
\begin{array}{rl}
 & Pfa \to Pfa \\ \hline
 & \neg\, Pa \,\&\, Pfa \to Pfa \\ \hline
 & \exists x\,(\neg\, Pa \,\&\, Pfa) \to Pfa \\ \hline
L'_2 & \neg\, Pfa,\ \exists x\,(\neg\, Pa \,\&\, Pfa) \to \\ \hline
L'_1 & \neg\, Pfa,\ \forall\beta\exists x\,(\neg\, P\beta \,\&\, Pf\beta) \to \\ \hline
 & \neg\, Pfa \,\&\, Pffa,\ \forall\beta\exists x\,(\neg\, P\beta \,\&\, Pf\beta) \to \\ \hline
E_0 & \exists x\,(\neg\, Pfa \,\&\, Pffa),\ \forall\beta\exists x\,(\neg\, P\beta \,\&\, Pf\beta) \to \\ \hline
 & \forall\beta\exists x\,(\neg\, P\beta \,\&\, Pf\beta),\ \forall\beta\exists x\,(\neg\, P\beta \,\&\, Pf\beta) \to \\ \hline
 & \forall\beta\exists x\,(\neg\, P\beta \,\&\, Pf\beta) \to
\end{array}
$$

After transformation into a proof satisfying Condition (A)

$$
\begin{array}{rl}
Pfa \to Pfa & \\ \hline
\neg\, Pa \,\&\, Pfa \to Pfa & l_1 = 0 \\ \hline
\neg\, Pfa,\ \neg\, Pa \,\&\, Pfa \to & l_2 = 3 \\ \hline
\neg\, Pfa,\ \exists x\,(\neg\, Pa \,\&\, Pfa) \to & \\ \hline
\neg\, Pfa,\ \forall\beta\exists x\,(\neg\, P\beta \,\&\, Pf\beta) \to & K_2 \text{ crit.} \\ \hline
\neg\, Pfa \,\&\, Pffa,\ \forall\beta\exists x\,(\neg\, P\beta \,\&\, Pf\beta) \to & \text{2-sing.} \\ \hline
\exists x\,(\neg\, Pfa \,\&\, Pffa),\ \forall\beta\exists x\,(\neg\, P\beta \,\&\, Pf\beta) \to & \text{2-sing.} \\ \hline
\forall\beta\exists x\,(\neg\, P\beta \,\&\, Pf\beta),\ \forall\beta\exists x\,(\neg\, P\beta \,\&\, Pf\beta) \to & \text{2-sing.} \\ \hline
\forall\beta\exists x\,(\neg\, P\beta \,\&\, Pf\beta) \to &
\end{array}
$$

Let us compare the height of D_1 with the height of D. Let us note that if π_1/π_2 is a pair of inferences of D which are different from L_1/L_2 and π'_1 corresponds in D_1 to the inference π_1 and π'_2 corresponds to π_2, then the collocation of π'_1 and π'_2 is the same as the collocation of π_1 and π_2. (More precisely, π_1 is higher than π_2 if and only if π'_1 is higher than π'_2.)

Let us prove first that i-singular pairs for $i < p$ do not emerge as a result of the permutation made. Let π'_1/π'_2 be an i-singular pair of D_1 and π_1/π_2 the corresponding pair of D. If π'_1/π'_2 does not belong to any of the types

$$E'_{11}/L'_2, \quad E'_{12}/(L'_2), \quad L'_2/L'_1, \quad (L'_2)/L'_1, \tag{7}$$

then π_1/π_2 is an i-singular pair in D, which, by virtue of the choice of the number p, implies that $i \geqslant p$. Let π'_1/π'_2 belong to one of the types (7).

If L'_1 is i-singular, then r_i is contained in r_p (since $L'_1 \in K_p$), whence $i > p$. If L'_1 is not an i-inference and $\pi'_2 = L'_2$ (i.e. π'_1/π'_2 belongs to one of the first two types of (7)), then π_1/L_2 is an i-singular pair in D, which (by virtue of the choice of p) implies $i \geqslant p$. If, finally, L'_1 is not an i-inference and $\pi'_2 = L'_1$, then there is an i-inference below L'_1. The pair consisting of L_2 and the inference in D which corresponds to that i-inference in D_1 is i-singular in D, whence again $i \geqslant p$. To prove that the height of D_1 is less than the height of the proof D, it remains to establish that $l_p(D_1) < l_p(D)$.

As already noted, to any i-singular pair of D_1 which does not belong to any of the types (7) there corresponds an i-singular pair in D which does not belong to any of the types

$$E_{11}/L_2,\ E_{12}/L_2,\ L_1/L_2. \tag{8}$$

Let us observe that to such distinct pairs of D_1 correspond distinct pairs of D, so that the "contribution" brought into $l_p(D_1)$ by pairs not belonging to any of the types (7) is not greater than the "contribution" brought into $l_p(D)$ by pairs not belonging to any of the types (8). It remains to evaluate the "contributions" brought by pairs of types (7) and (8). To any p-singular pair belonging to one of the first two types of (7) there corresponds a p-singular pair in the proof D, distinct pairs in D corresponding to such distinct pairs in D_1. ("New" p-singular pairs of these types in D_1, by comparison with those in D, could emerge only if L'_1 is a p-inference. However, $L'_1 \in K_p$, and therefore L'_1 cannot be a p-inference.) Therefore the "contribution" brought into $l_p(D_1)$ by pairs of the two types mentioned is not greater than the contribution brought into $l_p(D)$ by pairs of the types E_{11}/L_2 and E_{12}/L_2. Since L_1/L_2 is a p-singular pair in D, it remains for us to prove that neither of the pairs L'_2/L'_1 or $(L'_2)/L'_1$ is p-singular. This assertion follows from the fact that L_2 (and hence both L'_2 and (L'_2)) does not belong to the class K_p. (Otherwise L_1/L_2 would not be p-singular.) The discussion of Case 1 is finished.

Let us call a p-singular pair of adjacent inferences *extremal* if there are no p-singular pairs of adjacent inferences higher than it.

Case 2. Case 1 does not hold, and there is an extremal p-singular pair in D such that there are no inferences belonging to the class K_p above one of the premises of the lower member of this pair. Let us find one of the pairs satisfying the condition just formulated and call it the *distinguished premise*. Let us apply the notation of schema (5). Since L_2 is a two-premise inference, L_1/L_2 does not have any of the forms (4). Let us call the premise of L_2 above which L_1 is located the *first* premise, and the other premise of L_2 the *second* premise. Let us denote by V the descendant of the principal occurrence of L_1 in the conclusion of L_2, and V_1 the ancestor of V in the second premise.

Subcase 2.1. L_1/L_2 does not belong to type (3), or the occurrence V_1 is introduced into the second premise immediately by thinning. Let us represent L_1 and L_2. If L_1/L_2 has the form (3), the possibility of permuting follows from Lemma 8 of [2].

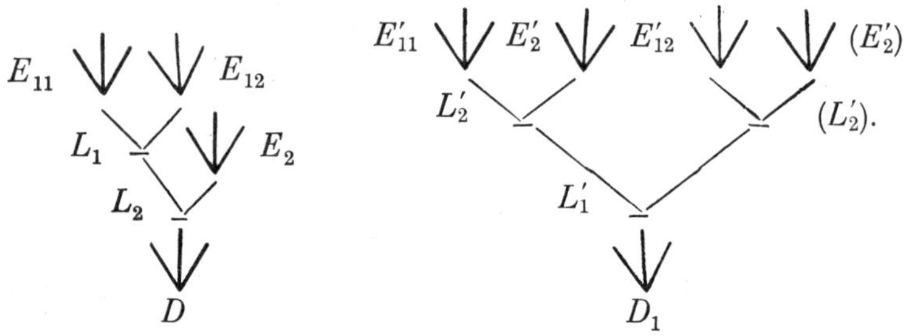

The assertion that the height of D_1 is less than the height of D will be proved analogously to the way the corresponding assertion in Case 1 was proved.

If π'_1/π'_2 is a pair of inferences of D_1 which does not belong to any of the types

$$E'_2/L'_1, \quad (E'_2)/L'_1, \quad L'_2/L'_1, \quad (L'_2)L'_1, \tag{9}$$

with π'_1 located higher than π'_2, and an inference π_1 of D corresponds to π'_1 and π_2 corresponds to π'_2, then π_1 is located higher than π_2.

Let π'_1/π'_2 be an i-singular pair ($i = 1, \cdots, m$). Let us prove that $i \geqslant p$. If π'_1/π'_2 does not belong to any of the types

$$E'_2/L'_1, \quad (E'_2)/L'_1, \tag{10}$$

this assertion is proved as in Case 1. If, however, $\pi'_1/\pi'_2 \in (10)$, then, since $L'_1 \in K_p$, this pair cannot be p-singular. Therefore pairs of the types (10) do not bring a "contribution" into $l_p(D_1)$. Utilizing this remark, it is not difficult to prove that $l_p(D_1) < l_p(D)$ by the same argument as in Case 1 by using the fact that E_2 and E'_2 do not contain inferences of K_p.

Subcase 2.2. L_1/L_2 belongs to type (3), and V_1 is traceable to an axiom. (Since D is a weeded inference, Case 2 is exhausted with this.) Let us show that in this case it is possible, without augmenting $l_p(D)$, without violating the condition on x-exceptional types, and preserving L_1/L_2 as an extremal pair, to ascertain that condition (b) of Lemma 8 of [2] would be fulfilled, i.e. that L_1 (which in our case is $\to \supset, \to \neg$ or $\to \&; L_1 \neq (\to \forall)$ in view of Condition (++)) is also above the second premise of L_2. We shall apply the transformation in the proof of the theorem on invertibility (cf. [4]) to the proof above the second premise of L_2. Reference to the same theorem is insufficient, since we must see to height. Let us consider the case when L_1 is $\to \&$. Let us transform the proof $E_{A \& B}$, which is above the second premise of L_2:

$$E_{A\&B}\left\{\begin{array}{c} \Downarrow \quad \Downarrow \\ \dfrac{\Sigma\to A \quad \Sigma\to B}{\Sigma\to A\,\&\,B} \\ \Downarrow \\ \Gamma\to A\,\&\,B \end{array}\right. \qquad E_2'\left\{\begin{array}{c} E_A\left\{\begin{array}{c}\Downarrow \\ \Sigma\to A \\ \Downarrow \\ \Gamma\to A\end{array}\right. \quad \left.\begin{array}{c}\Downarrow \\ \Sigma\to B \\ \Downarrow \\ \Gamma\to B\end{array}\right\}E_B \\ \hline \Gamma\to A\,\&\,B \end{array}\right.$$

E_A is obtained here from $E_{A\,\&B}$ by replacing all preimages of (the explicitly shown occurrence of) $A\ \&\ B$ by A and cancelling the right premises corresponding to $(\to\&)$; E_B is obtained analogously. The new proof E_2' contains exactly the same inferences as the old one but in a different order; therefore it does not contain inferences of K_i, $i \leqslant p$, so that l_p is not altered. Singular and quasisingular inferences of the x-exceptional types also do not appear; the inferences whose positions in the proof change under our transformation are of the same type as L_1, a quasisingular inference of the old proof.

The case $L_1 = \to\supset$ (as well as $L_1 = \to\neg$) is treated quite analogously. The corresponding transformation is

$$E_2\left\{\begin{array}{c} \Downarrow \\ \dfrac{A,\,\Sigma\to B}{\Sigma\to A\supset B} \\ \Downarrow \\ \Gamma\to A\supset B \end{array}\right. \qquad E_2'\left\{\begin{array}{c} \Downarrow \\ A\Sigma\to B \\ \Downarrow \\ \dfrac{A\Gamma\to B}{\Gamma\to A\supset B} \end{array}\right.$$

Hence it can be assumed that an inference by the same rule as L_1 is above the second premise of L_2. We denote it by (L_1). Let us carry out the permutation; i. e. let us transform (5) into (6). The restriction on x-exceptional types, and also the assertion that the height of D_1 is less than the height of D, are proved analogously to the way the corresponding assertion was proved in Case 1. The simplification, as compared to Case 2.1, consists in the fact that in D_1, as compared with D, only the pairs L_2'/L_1' and $(L_2')/L_1'$ turn out to be "invertible", while the collocation of pairs belonging to the types E_{21}'/L_1', E_{22}'/L_1' is preserved.

Case 3. Neither Case 1 nor Case 2 holds. Let us find some extremal pair, for example, the leftmost. Let us call the triple consisting of the upper and lower members of the pair found and a nonstructural rule application immediately above the second premise of the lower member of the pair found, the *distinguished system.* Let us apply the notation introduced in explanation of (5) (we denote by L_1 the upper member of the extremal pair found; in this connection (5) should be read to within the location of L_1 above one or the other of the premises of L_2). Let us prove that it is possible by cancelling certain sequent occurrences to ascertain that the third member of the distinguished system (let us denote it temporarily by L_3) will be an inference by the same rule as L_1, with the same principal and side formulas.

We first make some remarks supplementing Remarks 1–3 at the beginning of the proof.

REMARK 4. If Case 3 holds, both L_3 and all remaining inferences located between L_2 and the critical inferences of class K_p which are higher and to the right of L_2 (and there are such by virtue of the condition defining Case 3) belong to class K_p; otherwise the p-singular pair found would not be extremal.

REMARK 5. Let some sequent S of D be of the form $S_1 \cdot S_2 \cdot T$, where S_1 and S_2 are one-member sequents belonging to the class K_i, i.e. there are inferences π_1 and π_2 of class K_i whose principal formulas belong[10] respectively to (the explicitly shown occurrences of) S_1 and S_2. In addition, let the image (of the only formula occurring in S_1) in the bottom sequent of the proof D be contained in the image of S_2, and let $S_1 \neq S_2$. Then the following assertions are true:

1) S_2 is a sequent of the form $(B \supset C) \rightarrow$, and above the left premise of π_2 there is an inference π'_1, with principal formula S_1, which belongs to (a side formula of) π_2, all inferences transforming the principal formula of π'_1 into a side formula of π_2 being succedent sequents different from $\rightarrow \supset$ and $\rightarrow \neg$.

2) S_1 is a purely succedent sequent, i.e. of the form $\rightarrow F$.

As a matter of fact, if the images of S_1 and S_2 coincide, then $S_1 = S_2$ by virtue of Lemma 3.1. If, however, the image of S_1 is contained strictly within the image of S_2, below (the occurrence under consideration of) the sequent S there is an inference κ_2 with principal formula S_2, S_1 belonging to a side formula of κ_2; if the proof D is looked at from below upwards, κ_2 is one of the inferences "splitting" the descendant of S_1 in the bottom sequent up to S_1; in the process of such a splitting it is necessary to pass through a formula with the same image as S_2. If κ_2 is different from $\supset \rightarrow$ or κ_2 is $\supset \rightarrow$ but S is located above its right premise, i.e. above an antecedent side formula, then π_2 is a semidegenerate inference, contradicting Condition (A). Therefore S is located above a succedent side formula of κ_2, i.e. exactly the latter formula is the descendant of S_1. We find above π_2 the inference π'_1, with principal formula S_1 belonging to a side formula of π_2. (Such an inference is necessary in order to split S_2 up to the principal formula of a hypercritical inference.) Let us prove that $\rightarrow \supset$ and $\rightarrow \neg$ are not among the inferences transforming (the principal formula of) π'_1 into (a side formula of) π_2. Indeed, if such a $(\rightarrow \supset)$- or $(\rightarrow \neg)$-inference were found, it would be semidegenerate: an inference by the same rule with the same principal formula would transform S_1 into the principal formula

(10) We identify a one-member sequent with its only term.

of κ_2. Therefore the sequent S_1 cannot be purely antecedent; for a succedent side formula of the inference π_2 to have an antecedent ancestor, it is necessary that a $(\to\supset)$- or $(\to\neg)$-inference be among them.

Returning to consideration of Case 3, let us prove that L_3 can be cancelled. By the same token $l_p(D)$ is diminished and the induction step will be justified. The part of the proof in which we are interested has, by virtue of Remark 5, the form

$$\dfrac{L_1\dfrac{\Downarrow}{\Sigma_1 \to B'} \qquad \dfrac{F\left\{\begin{array}{c}\dfrac{\Downarrow}{\Sigma' \to B'}\,\pi'_1\\ \Downarrow\\ \Sigma'_2 \to B\end{array}\right. \qquad \begin{array}{c}\Downarrow\\ \Sigma''_2 C \to G\end{array}}{(B\supset C)\,\Sigma_2 \to G}\,L_3}{(B\supset C)\,\Sigma \to B'}\,L_2,$$

Let us prove that $\Sigma' \subset \Sigma_2$. Since $\pi'_1 \in K_p$, all nonstructural inferences of F belong, by virtue of Remark 3, to K_p. By virtue of Remark 4, there are no antecedent inferences among them: antecedent inferences of the class K_p could be only $(\supset\to)$-inferences, and in that case (an explicitly shown occurrence of a formula of) B' would not be an ancestor of B. Therefore only structural inferences and $\to\&$, $\to\vee$, $\to\exists$ and $\to\forall$ can be encountered in F. All of these inferences do not augment the antecedent. Now F and L_3 can be cancelled; if G is B', we obtain

$$\dfrac{\Sigma_1 \to B' \qquad \dfrac{\Sigma' \to B'}{(B\supset C)\,\Sigma_2 \to B'}}{(B\supset C)\,\Sigma \to B'}\,.$$

If $G \neq B'$, then L_2 is $(\supset\to)$, and Σ_2 is contained in Σ;

$$\dfrac{\dfrac{\Sigma' \to B'}{\Sigma_2 \to B'}}{(B\supset C)\,\Sigma \to B'}\,.$$

The proof of the theorem is completed.

§5. Syntactical sufficient conditions for "regularizability"

Theorem 5.2, which is proved in this section, gives a syntactical sufficient condition for it to be possible to reconstruct every proof of the sequent $\mathfrak{S}(S, x)$ into a proof satisfying the condition of Theorem 4.1. For later applications we need a theorem on "global" permutation for deductions (which are not in general proofs), i.e. one of the possible generalizations of Theorem 2 of [2] to deductions. Let us call a deduction *weeded* if it possesses the property formulated in Lemma 4 of [2], i.e. if thinnings can occur in it only as follows: a) immediately preceding the endsequent; b) introducing a side formula of a $(\to\supset)$-inference whose second side formula is not introduced by a thinning; c) introducing one of the parametric formulas into a premise of a two-premise inference (but not into the succedent of an $\supset\to$), the corresponding occurrence in the other premise not being introduced by a thinning.

It is obvious that the bound part of a weeded deduction is a weeded deduction.

THEOREM 5.1. *Let* D *be a pure weeded deduction of a purified sequent* T, *and let a break-down into the classes* $C_1, \cdots, C_k$ *of the occurrences of logical symbols in* T *which are split*(11) *in* D (C_i *is said to be higher than* C_j *if* $i < j$) *be chosen which meets the following restrictions*:

a) *If a logical symbol occurrence* L_2 *dominates the occurrence* L_1, *then* L_1 *is in the same class as* L_2 *or in a higher one.*

b) *If the type of the occurrence* L_1 *is located above the line, and the type of* L_2 *is located under the line, in the list cited below, then* L_2 *is not in a higher class than* L_1:

$$\frac{\supset\rightarrow,\ \neg\rightarrow}{\rightarrow\supset,\ \rightarrow\neg} \qquad \frac{\rightarrow\supset,\ \supset\rightarrow,\ \rightarrow\&,\ \rightarrow\vee,\ \rightarrow\neg,\ \neg\rightarrow,\ \rightarrow\forall,\ \rightarrow\exists}{\vee\rightarrow}.$$

c) *If a negative quantifier of* L_1 *gets into a lower class than a positive quantifier of* L_2, *either the characteristic term of any* L_1*-inference in* D *does not contain the characteristic variable of any* L_2*-inference or* L_1 *is* $\rightarrow\exists$ *and* L_2 *is* $\rightarrow\forall$.

Then the applications of logical rules in this deduction can be permuted so that a deduction of the sequent T *is obtained from the old original sequents in which no inference belonging to an occurrence of a higher class is below an inference belonging to a lower class.*

REMARK 1. We are thus considering an incomplete classification which does not include symbols not subject to splitting.

The proof of this theorem is easily obtained by simplifying the proof of Theorem 2 of [2], and we shall not carry it out. Our additional restrictions make it possible not to apply Lemmas 8 and 9 of [2] and to restrict ourselves to Lemma 7, whose application does not alter the original sequents, as well as to the remark on the possibility of permuting a minus-inference and a plus-inference if the characteristic term of the first does not contain the characteristic variable of the second.

THEOREM 5.2. *Let* S *and* x *satisfy Condition* (++), *and let the fact that some occurrence of one of the types* $\neg\rightarrow$ *or* $\supset\rightarrow$ *(respectively, of one of the types* $\rightarrow\vee$ *or* $\rightarrow\exists$*) dominates the quantifier* Qx *imply that no occurrence of the type* $\vee\rightarrow, \rightarrow\neg$ *or* $\rightarrow\supset$ *(respectively, of the type* $\vee\rightarrow$*) is within the scope of a negative quantifier*(12) *(respectively, within the scope of a negative* $\forall$*-quantifier). Then with respect to every proof of the sequent* $\S(S, x)$ *it is possible to construct a proof of the same sequent which satisfies the condition of Theorem* 4.1.

---

(11) We say that an occurrence of a logical symbol is split in D if there is an inference in D which belongs to that occurrence.

(12) It is easy to see the connection of these conditions with the pairs (2), §4.

PROOF. Let S and x satisfy the condition and D be a proof of $\mathcal{S}(S, x)$. By virtue of Theorem 2.1 it can be assumed that D satisfies Condition (A). Let us suppose that an occurrence of at least one of the types $\supset\rightarrow$, $\neg\rightarrow$ or $\vee\rightarrow$ dominates the quantifier Qx. (Otherwise, the conditions of Theorem 4.1 are satisfied trivially.)

Case 1. Occurrences of the types $\neg\rightarrow$ or $\supset\rightarrow$ dominate the quantifier Qx. In this case, in view of Condition (++), Q is $\exists$. Preparatory to applying Theorem 2 of [2], let us break down all logical symbol occurrenees in $\mathcal{S}(S, x)$ into two classes. (This break-down is connected in an obvious way with the restriction on occurrences of logical symbols in S which is imposed in the condition.) Let us include in the first (the higher, in the terminology of Theorem 2 of [2]) class C_1: a) all occurrences of the types $\rightarrow\exists$ and $\forall\rightarrow$; b) all logical symbol occurrences within the scope of the occurrences already included in the first class. Let us include in the second (lower) class C_2 all remaining occurrences. The conditions of Theorem 2 of [2] are met by this break-down. Indeed, the proof D satisfies Condition (A); therefore it is pure and weeded, i. e. satisfies the condition of Lemma 4 of [2]. Fulfillment of item a) of the condition of Theorem 2 follows from item c) of the definition of class C_1. It remains to prove that no pair of inferences L_1/L_2, where $L_1 \in C_2$ and $L_2 \in C_1$, belongs to any of the types §4 (2) or §4 (4). The fact that no pair of the indicated kind belongs to the type §4 (4) follows from the fact that all occurrences of the types $\rightarrow\exists$ and $\forall\rightarrow$ belong to the upper class. The absence of "bad" pairs of the type §4 (2) follows from the fact that by virtue of the theorem's condition, occurrences of the types $\vee\rightarrow$, $\rightarrow\neg$ and $\rightarrow\supset$ do not get into the upper class.

It is now possible by virtue of Theorem 2 of [2] to construct a proof of $\mathcal{S}(S, x)$ in which no inference of C_2 is above any inference of C_1. By applying the proof of Theorem 1.2 to the constructed proof it is possible to establish fulfillment of Condition (A). The constructed proof satisfies the condition of Theorem 4.1, since inferences of the x-exceptional types belong to the lower class and cannot be above minus-inferences.

Case 2. Qx is not within the scope of $\supset\rightarrow$ or $\neg\rightarrow$. In this case the only x-exceptional type is $\vee\rightarrow$. Let us show how to get rid of singular inferences of this type. Let us say that an inference π belongs to the class K_i^* $(i = 1, \cdots, m)$ if it belongs to K_i and is a succedent inference.

REMARK 2. The fact that Qx is not within the scope of $\supset\rightarrow$ or $\neg\rightarrow$ implies that any inference located below some inference of K_i^* also belongs to K_i^*. (More precisely, a side formula of the lower inference is the descendant of the principal formula of the upper inference.)

REMARK 3. By replacing some f-term occurrences with variables it is possible to establish that i-inferences by the rule $\forall\rightarrow$ are the only minus-inferences whose characteristic terms contain f and which are below some inferences of K_i^* $(i = 1, \cdots, m)$ or belong to K_i^*. (We shall assume in the sequel that such replacements have already been made.)

As a matter of fact, if this condition is violated for some inference π, it is possible using Lemma 2 of §4 of [5] to replace the characteristic term t of the inference π with a new term not containing the functor f. The conditions of the lemma mentioned will be fulfilled for all inferences higher than π, since, by virtue of Remark 1, all quasiterms which can be candidates for the role of r' of item c) of this lemma contain t.

Subcase 2.1. In the given proof inferences of K_i^* $(i = 1, \cdots, m)$ are not the upper members of singular pairs.

2.1.1. Q is $\forall$. The proof under consideration generally does not contain singular pairs.

2.1.1. Q is $\exists$. Let us consider the uppermost inferences belonging to the classes K_i^*. Let T be a premise of one of them which contains Qx. Let us prove that it is possible to remove singular inferences from the proof above T (and by the same token, by virtue of Remark 3, from the whole of D). Let us write T in the form $F, \Gamma \to \Delta$, where F is a formula containing $\exists x$. Let us show how to transform T into a sequent of the form $\mathfrak{S}(T_1, x)$. Since it can be assumed, by virtue of Remark 3, that below T there are no minus-inferences removing f, it can therefore be assumed that all occurrences of f in T are within F and are traced to occurrences of f in the bottom sequent of the proof D. More precisely, F has the form $[H]^{\beta_{1 \div l}}_{s_{1 \div l}}$, $l \leqslant k$, where H is the image of F in the bottom sequent, the terms $s_1, \cdots, s_l$ not containing f. Here $\beta_1, \cdots, \beta_l$ are those of the variables $\beta_1, \cdots, \beta_k$ which do not occur bound in H. H itself is obtained from some subformula M of the sequent S (we are considering a proof of the sequent $\mathfrak{S}(S, x)$) on replacing $\exists x A$ by $\exists x [A]^x_{f(\beta_{1 \div k})}$. Therefore F is obtained from $[M]^{\beta_{1 \div l}}_{s_{1 \div l}}$ on replacing $\exists x B$ (where B is $[A]^{\beta_1 \ldots \beta_l}_{s_1 \ldots s_l}$) by $\exists x [B]^x_{f(s_{1 \div l}, \beta_{l+1 \div k})}$. Now let us consider the sequent T_1

$$[M]^{\beta_{1 \div l}}_{s_{1 \div l}}, \quad \Gamma \longrightarrow \Delta.$$

The sequent $\mathfrak{S}(T_1, x)$ differs from T only by the replacement of $f(s_{1 \div l}, \beta_{l+1}, \cdots, \beta_k)$ by $g(\beta_{l+1}, \cdots, \beta_k)$. Let us perform the same operation (replacing f by g and cancelling the first l arguments) over the whole proof of the sequent T. The figure obtained will be a proof of $\mathfrak{S}(T_1, x)$ which satisfies the condition of Theorem 4.1. (The occurrence of $\exists x$ in T_1 can be dominated only by the connectives $\vee, \&, \forall$.) Applying Theorem 4.1 to it and then making the reverse substitution of $f(s_{1 \div l}, t_{l+1}, \cdots, t_k)$ for $g(t_{l+1}, \cdots, t_k)$, we obtain a proof which generally does not contain singular inferences.

Subcase 2.2. There are inferences of K_i^* $(i = 1, \cdots, n)$ which are members of singular pairs. By virtue of the stipulation made after Remark 3, we obtain the result that below some inferences of K_i^* there are i-inferences by the rule $\forall \to$. Let us show how to get rid of such i-inferences. (By the same token everything will reduce to Case 2.1, and the theorem will be proved.) Let us consider the part of the given proof which is the complement of the union of parts considered in Subcase 2.1.1. This is the deduction (let us denote it by E) whose original sequents are the premises of the uppermost inferences of K_i^*. We may assume that an i-inference by the rule $\forall \to$ is the last inference in it. (Lifting i-inferences above K_i^* will not affect the parts of the original deduction which are below all i-inferences.)

Preparatory to applying Theorem 5.1, let us break down into the two classes C_1 and C_2 all occurrences of logical symbols in the bottom sequent of E which are split in E. Let us attribute to the upper class C_1 all occurrences of the type $\forall \to$ and all occurrences split in E and dominated by occurrences of C_1; let us attribute to the lower class C_2 all remaining occurrences split in E. Let us verify fulfillment of the conditions of Theorem 5.1. Condition a) is fulfilled by virtue of the definition of C_1. Condition b) is ensured by the conditions defining Case 2: occurrences of the type $\forall \to$ generally do not dominate occurrences of the type $\vee \to$; occurrences of the types $\to \supset$ and $\to \neg$, however, which are dominated by occurrences of the type $\forall \to$ are not split in E; by virtue of Remark 1 the corresponding inferences would belong to the class K_i^*, and hence an $(\forall \to)$-inference would belong to the class K_i, which would imply that Qx is within the scope of $\supset \to$ or $\neg \to$.

Let us verify the fulfillment of condition c). Let some minus-inference L_2 belong to C_2. Then L_2 is $\to \exists$ (all $\forall \to$ are put into C_1) and belongs to K_i^* (Remark 1). If the characteristic variable b of some plus-inference L_1 is contained in the characteristic term of L_2, b is contained in r_i (as the term belonging to critical figures of the class K_i was denoted in §3). In that case our deduction (and a fortiori the whole given proof) would not possess the property of purity of variables: the lower $\forall \to$ contains r_i (it is an i-inference) and hence also b, which is impossible. Applying Theorem 5.1, we lift all $\forall \to$ above K_i^*, which was required.

§6. Justification of the Skolem method of eliminating quantifiers

Theorem 5.2 will be utilized in this section to justify the admissibility of applying the Skolem method of eliminating quantifiers (SME) in certain cases.

THEOREM 6.1. *Let x and S satisfy Condition* (+) *and all conditions of Theorem* 5.2 *except perhaps Condition* (++). *Then the provability of the sequent $\mathfrak{S}(S, x)$ implies the provability of the sequent S, i. e., applying SME to S and x is admissible.*

PROOF. We prove the theorem by induction on the number of S-positive quantifiers dominating the quantifier Qx.

Induction basis. S and x satisfy all conditions of Theorem 5.2. Let D be an arbitrary proof of the sequent $\mathfrak{S}(S, x)$. By virtue of Theorem 1.2 it can be assumed that D satisfies Condition (A). By virtue of Theorems 5.2 and 4.1 it can be assumed that there are no singular pairs of rule applications in D. This, Theorem 3.2 and the fact that D satisfies Condition (A) imply that in D all critical rule applications are regular. By virtue of Theorem 2.1 this implies that it is possible to construct a proof of S.

Let us justify the induction step. With the help of the Skolem method we remove the leftmost positive quantifier dominating the quantifier Qx, and then restore (utilizing the induction hypothesis) the quantifier Qx itself and the first of the quantifiers removed. The theorem is proved.

We extract a corollary connected with the admissibility of "Skolemizing" axiomatic theories based on the intuitionistic predicate calculus.

THEOREM 6.2. *Let S be a purified sequent and $\exists x$ an S-positive quantifier not within the scope of the symbols $\supset$ and $\neg$. Then applying SME to S and x is admissible.*

A special case of Theorem 6.2 is

THEOREM 6.3. *The rule*

$$\frac{\forall x_1 \ldots \forall x_n [A]^y_{f(x_{1 \div n})} \supset B}{\forall x_1 \ldots \forall x_n \exists y A \supset B},$$

where f does not occur in A or B and $x_1, \cdots, x_n$ are free for y in A, is admissible in the intuitionistic predicate calculus.

§7. Applying SME to sequents with empty succedents

As is well known, an elementary axiomatic theory with axioms $A_1, \cdots, A_k$ is consistent if and only if the sequent $A_1, \cdots, A_k \rightarrow$ is unprovable. Of interest in connection with this is the question of which of the unprovable sequents with empty succedent remain unprovable after applying SME. Apparently it is difficult to expect applicability of SME to positive antecedent $\forall$-quantifiers. Therefore in the theorems cited below it will be a question of the admissibility of applying SME to $\exists$-quantifiers.

THEOREM 7.1. *Let S be a purified sequent with empty succedent and not containing occurrences of the type $(\rightarrow \forall)$. Then applying SME to any S-positive quantifier is admissible.*

This assertion follows from the fact that classical provability of sequents with empty succedent and not containing occurrences of the type $(\rightarrow \forall)$ is equivalent to their intuitionistic provability (cf. [6], Theorem 3).

THEOREM 7.2. *Let S be a purified sequent with empty succedent, and let $\exists x$*

be an S-positive quantifier which perhaps only propositional connectives and occurrences of the type $\exists\rightarrow$ *dominate. Then applying SME to the pair* S, x *is admissible.*

The theorem is proved by induction on the number of positive quantifiers dominating the quantifier under consideration. The induction step is justified exactly as in Theorem 6.1.

Let us justify the induction basis. Let us observe first that SME is applicable to an outer occurrence of the existential quantifier: the sequent $\exists xA,\ \Gamma\rightarrow\Delta$ is provable if and only if $[A]^x_f,\ \Gamma\rightarrow\Delta$ is provable. Since the usual rules for taking $\exists$ out from under $\vee$ and $\&$ are valid in the intuitionistic predicate calculus, SME is applicable to positive $\exists$-quantifiers not within the scope of $\forall, \supset$ or $\neg$. Finally, in our case (empty succedent and absence of $\forall$-occurrences dominating the given $\exists$) it is possible, by using Glivenko's theorem to exclude $\supset$ and bring $\neg$ inside, to apply SME, and then by again using Glivenko's theorem to restore $\supset$ and $\neg$.

§8. Necessity for the restrictions introduced

It will be shown in this section that all restrictions on the pair S, x in the condition of Theorem 6.1 are actually necessary in the sense that with violation of at least one of them applying SME to the pair S, x may turn out to be inadmissible. An example showing the inadmissibility of applying SME to a positive quantifier dominated by an occurrence of the type α when there is an occurrence of the type β within the scope of a negative Q-quantifier will be designated by the symbol $\alpha/Q\beta$. In each of the counterexamples cited below a purified sequent S will be indicated which possesses the following properties:

1) S is unprovable in the intuitionistic predicate calculus.

2) There is an S-positive quantifier $\exists x$ satisfying Condition (+); moreover, if the succedent of S is not empty, no quantifiers dominate this quantifier; if the succedent of S is empty, exactly one (and negative, at that) quantifier dominates $\exists x$.

3) The result of applying SME to the pair (S, x) is provable in the intuitionistic predicate calculus.

Let us observe that by virtue of Theorem 7.1 each counterexample with empty succedent must contain $(\rightarrow\forall)$, and, by virtue of Theorem 7.2, x must be dominated by a negative quantifier (provided Condition (+) is not violated). Without great difficulty the reader can verify the presence of property (3) in each concrete case.

The note, "cf. cells $\alpha_1/Q\beta_1, \alpha_2/Q\beta_2$," in a line of the table signifies that any unprovable sequent satisfying the condition defining the line under consideration satisfies one of the conditions $\alpha_1/Q\beta_1$ or $\alpha_2/Q\beta_2$, as well.

There are counterexamples of the elimination of $\forall$-quantifiers in the last two cells of the table.

Let us prove the unprovability of the sequents listed in the table.

Cell Number	Restriction	Counterexample
1	2	3
1	$\neg\to/\forall\ \vee\to$	$\neg\neg\,\exists xFx,\ \forall y\,(Fy \vee p)\to p$
2	$\neg\to/\exists\ \vee\to$	Cf. cells $\neg\to/\exists\,(\to\neg)$, $\neg\to/\exists\,(\to\supset)$
3.1	$\neg\to/\forall\,(\to\neg)$	$\neg\neg\,\exists xFx,\ \forall y\,(\neg\neg\,Fy\supset p)\to p$
3.2		$\forall u\,\neg\neg\,\exists xFux,\ \neg\,\forall u\,(\forall y\,(\neg\neg\,Fuy\supset Pu)\supset Pu)\to$
4	$\neg\to/\exists\,(\to\neg)$	Replace $\forall y(\neg\neg$ in cell 3 by $(\exists y\,\neg\neg$
5	$\neg\to/\forall\,(\to\supset)$	Replace occurrences of the type $\to\neg$ in cell 3 which are within the scope of $\forall y$ in accordance with the equivalence $\neg B=(B\supset\pi)$
6	$\neg\to/\exists\,(\to\supset)$	Replace $\forall y(\neg\neg$ in cell 3 by $(\exists y\,\neg\neg$
7	$\supset_1\to/\forall\ \vee\to$	$(\exists xFx\supset q)\supset q,\ \forall y\,((Fy\supset q)\vee p)\to p\vee q$ empty succedent: cf. cells $\supset_1/\forall\,(\to\neg)$, $\supset_1/\forall\,(\to\supset)$
8	$\supset\to/\exists\ \vee\to$	Cf. cells $\supset\to/\exists\,(\to\neg)$, $\supset\to/\exists\,(\to\supset)$
9	$\supset_1\to/\forall\,(\to\neg)$	Utilize cell 3 analogously to the way square 1 is used in cell 7
10	$\supset_1\to/\exists\,(\to\neg)$	Replace $\neg\neg\exists xFx$ in cell 6 by $(\neg\exists xFx\supset\pi)$ and $\neg\neg\exists xFux$ by $\neg(\exists xFux\supset\pi)$
11	$\supset_1/\forall\,(\to\supset)$	$(\exists xFx\supset q)\supset q,\ \forall y\,(((Fy\supset q)\supset q)\supset p)\to p$ empty succedent: replace $\to\neg$ in cell 9 by $\to\supset$
12	$\supset_1/\exists\,(\to\supset)$	Replace $\to\neg$ in cell 10 by $\to\supset$
13.1	$\supset_2\to/\forall\ \vee\to$	$p\supset\exists xFx,\ \forall y\,((Fy\supset s)\vee(p\supset s)\vee q)\to(p\supset s)\vee q$
13.2		$\forall u\,(pu\supset\exists xFux),\ \neg\,\forall u\,(\forall y\,(\neg\,Fuy\vee\neg\,pu\vee qu)\supset\neg\,pu\vee qu)\to$
14.1	$\supset_2/\forall\,(\to\neg)$	$q\supset\exists xFx,\ \neg\neg\,q,\ \forall y\,(\neg\neg\,Fy\supset p)\to p$
14.2		$\forall u\,(qu\supset\exists xFux),\ \forall u\,\neg\neg\,qu,\ \neg\forall u\,(\forall y\,(\neg\neg\,Fuy\supset pu)\supset pu)\to$
15	$\supset_2/\exists\,(\to\neg)$	Obtained from cell 14 by bringing $\forall y$ inside the implication
16.1	$\supset_2/\forall\,(\to\supset)$	$q\supset\exists xFx,\ (q\supset s)\supset s,\ \forall y\,(((Fy\supset s)\supset s)\supset p)\to p$
16.2		empty succedent: replace $\neg Fy$ in cell 9 by $(Fy\supset\pi)$
17	$\supset_2/\exists\,(\to\supset)$	Obtained from line 16 by bringing $\forall y$ inside the implication
18	$\to\vee/\forall\ \vee\to$	$\forall y\,(p\vee(Fy\supset q))\to p\vee(\exists xFx\supset q)$; $\forall y\,(p\vee(Fy\supset q))\to p\vee\forall x\,(Fx\supset q)$ Empty succedent: cf. cells $\neg\to/\forall\vee\to$, $\supset\to/\forall\vee\to$
19	$\to\exists/\forall\ \vee\to$	$\forall u\,\forall v\,(Fau\vee Fbv)\to\exists y\forall xFyx$ Empty succedent: cf. cells $\neg\to/\forall\vee\to$, $\supset\to/\forall\vee\to$

Cell 1. By results of Kripke [7] the figure

$$\begin{array}{l}\{a,\ b\}\ [Fb,\ p]\\ \vert\\ \{a\}\end{array}$$

is a countermodel(13) for the sequent under consideration. The list of formulas having the value "true" (at a corresponding "moment of time") is shown in square brackets.

Cell 3.1. Implies line 1.

---

(13) The use of constructively defined countermodels to prove unprovability is justified entirely constructively in [7].

Cell 3.2. It is sufficient to establish the unprovability of the sequent

$$\forall u \neg\neg \exists x Fux, \quad \neg \forall u \exists y \neg\neg Fuy \rightarrow,$$

obtained (to within equivalence) from line 3.2 by replacing Pu with $\exists y \neg\neg Fuy$. The model

$$\vdots$$
$$\{a_1, \ldots, a_i, a_{i+1}\}\ [Fa_1a_2, \ldots, Fa_{i-1}a_i, Fa_ia_{i+1}]$$
$$\{a_1, \ldots, a_i\}\ [Fa_1a_2, Fa_2a_3, \ldots, Fa_{i-1}a_i]$$
$$\vdots$$
$$\{a_1, a_2\}\ [Fa_1a_2]$$
$$\{a_1\}.$$

serves as a countermodel for this sequent.

Cells 4–6. The sequents are equivalent to the sequents of Cell 3.

Cell 7. Substituting $q = \pi$, we obtain Cell 1.

Cell 11. Substituting $q = \pi$, we obtain Cell 3.1.

Cell 13.1. 13.2 is obtained from 13.1 by substitution.

Cell 13.2. Countermodel:

$$\vdots$$
$$\{a_1, b_1, \ldots, a_i, b_i, a_{i+1}\}\ [Pa_i, Qa_i, Fa_ib_i]$$
$$\vdots$$
$$\{a_1, b_1, a_2, b_2, a_3\}\quad [Pa_2, Qa_2, Fa_2b_2]$$
$$\{a_1, b_1, a_2\}\quad [Pa_1, Qa_1, Fa_1b_1]$$
$$\{a_1\}.$$

Cell 14.2. Substituting $Qu = \neg\neg\exists x Fux$, we obtain 3.2.

Cell 14.1. 14.2 is obtained from 14.1 by substitution.

Cell 16.1. $s = \pi$ gives 14.1.

Cell 18. Countermodel:

$$\{a, b\}\ [p, Fb]$$
$$\{a\}.$$

Cell 19. Countermodel:

$$[Fa,a, Fab, Fac]\ \{abc\} \qquad \{a, b, d\}\ [Fba, Fbb, Fbd]$$
$$\{a, b\}\ [Faa, Fab].$$

§9. SME in positive and minimal calculi

As is well known, the positive predicate calculus is obtained by removing from the language of the intuitionistic predicate calculus the symbol $\neg$, and from the list of postulates those in which this symbol is used. A sequent not containing the symbol $\neg$ is provable in the positive predicate calculus if and only if it is provable in the intuitionistic one. In this section the expression "applying SME to the pair S, x is admissible" will signify that the sequent S is provable in the positive predicate calculus if and only if the sequent $\mathcal{S}(S, x)$ is provable in that calculus. The preceding implies that the assertion obtained if everything relevant to the symbol $\neg$ is eliminated from the formulation of Theorem 6.1 and the words "applying SME is admissible" are understood in the new sense is valid.

Counterexamples showing that the indicated assertions cannot be essentially weakened are contained in §8; the sequents cited there which correspond to cases in which the symbol $\neg$ does not appear do not contain this symbol.

The question of the admissibility (in the new sense) of applying SME to sequents with empty succedent is decided trivially: it is easy to see that no sequent with empty succedent is provable in the positive predicate calculus.

As to the minimal predicate calculus, as is well known, it is reduced to the positive one by replacing the constant π with a new variable p and subsequent replacement of formulas of the form $\neg A$ by $A \supset p$. Using this remark, it is possible to decide the question of the applicability of SME to the minimal calculus.

§10. ϵ-theorems in the intuitionistic predicate calculus

The question of the admissibility of the Skolem method of eliminating quantifiers was posed by H. Rasiowa in connection with the question of carrying over the Hilbert ϵ-theorems to this calculus.

Glivenko's theorem suggests that the first ϵ-theorem (the theorem on preserving the consistency of quantifier-free theories) should carry over to the constructive case, while examples in the preceding sections suggest that the second ϵ-theorem should not. Such is the situation in reality.

Let us denote by LJ^{ϵ} the calculus obtained by adding the ϵ-symbol to the language of the calculus LJ and the axiom schema

$$[A]^x_r \supset [A]^x_{\varepsilon_x A}, \tag{1}$$

where r is an arbitrary term, to its axiomatics.

THEOREM 10.1. *Let* Γ *be an arbitrary list of prenex formulas not containing the symbol* ϵ, *and let the sequent* $\Gamma \rightarrow$ *be provable in* LJ^{ϵ}. *Then it is provable in* LJ.

PROOF. By virtue of the second ϵ-theorem (for the classical predicate calculus), the provability of $\Gamma \rightarrow$ in LJ^{ϵ} implies its provability in LK. Its provability in LK implies (by virtue of Theorem 3 of [6]) its provability in LJ, which was required.

This theorem implies the analogue of the first ϵ-theorem for the intuitionistic predicate calculus (and even for the predicate calculus with equality: indeed, the axioms for equality are prenex formulas).

However, the second ϵ-theorem is false for LJ, even as a theorem on preserving consistency, since the formula

$$\neg\neg\forall u \exists x (\neg Fux \supset (\neg Fua \& \neg Fub)),$$

which is provable in LJ^{ϵ} with the help of the axioms $Fua \supset Fu\epsilon_x Eux$ and $Fub \supset Fu\epsilon_x Eux$ is not provable in LJ, which is easy to establish with the help of Gentzen methods.

The generalization of the first ϵ-theorem (trivial in the classical case) consisting in the fact that adding ϵ-axioms to a quantifier-free theory does not augment its extension, also does not carry over to the constructive case. Thus, for example, the formula p is unprovable by means of the calculus LJ from the formula

$$\forall x((\neg Fx \supset (\neg Ff \& \neg Fg)) \supset p)$$

(which is obvious if the formula $\exists x(\neg Fx \supset (\neg Ff \& \neg Fg))$ is substituted for p; however, it becomes provable if the axioms

$$Ff \supset F\varepsilon_x Fx, \quad Fg \supset F\varepsilon_x Fx$$

are added).

§11. SME in a predicate calculus with equality

Let us denote by $LJ^{=}$ the result of adding the two-place predicate symbol $=$ to the language LJ, and the axiom schema $\rightarrow r = r$ and the two rules for equality

$$\frac{[\Gamma]^a_r \rightarrow [\Delta]^a_r}{r = s,\ [\Gamma]^a_s \rightarrow [\Delta]^a_s}, \quad \frac{[\Gamma]^a_s \rightarrow [\Delta]^a_s}{r = s,\ [\Gamma]^a_r \rightarrow [\Delta]^a_r}$$

to the aggregate of its postulates.

It is very well known that the cut can be eliminated from $LJ^{=}$. The addition of equality has roughly the same impact on the admissibility of SME as addition of the ϵ-symbol.

It is unconditionally applicable to sequents of the form $\Gamma \rightarrow$, where Γ is a list of prenex formulas (Theorem 10.1), but there is a counterexample of the form $\Gamma \rightarrow p$: the sequent

$$\forall y \exists x([\neg(Gx \& y = a) \supset Fy] \& [(\neg Gy \supset (Fa \& Fb)) \supset p]) \rightarrow p \tag{1}$$

is unprovable in $LJ^{=}$ (the proof of this assertion being analogous to the proof below of the unprovability of sequent (2)), while the result of applying SME to it is provable in $LJ^{=}$. What is more, as the example of the sequent

$$\forall y \forall u \exists x(\neg(Gux \& y = u) \supset Fy),\ \neg\forall w v \exists z(\neg Gvz \supset (Fv \& Fw)) \rightarrow \tag{2}$$

shows, even applying SME to the quantifier $\exists x$ in an antecedent prenex formula (if the other antecedent members are not prenex formulas) can convert a consistent axiomatic theory into an inconsistent one.

Let us prove the unprovability of sequent (2). Let us consider an arbitrary nondegenerate proof D of (2), and let $\pi_\forall$ be the uppermost $(\rightarrow \forall)$-inference in D, and π_E the uppermost $(\to \exists)$-inference. In view of the weededness of D, the principal formula of π_E belongs to the principal formula of $\pi_\forall$. By virtue of this, only antecedent inferences by the rules $\forall \to$ and $\exists \to$ between π_E and $\pi_\forall$ can occur, since all remaining possible inferences (if they are examined "from below upwards") annihilate the succedent in at least one of the branches. In view of the reversibility of the rules $\to \forall$, $\to \supset$ and $\to \&$, it can be assumed that the section interesting us has (to within thinning) the form

$$\dfrac{\dfrac{\dfrac{(S_1)\ \Gamma\Sigma\neg Gvz \to Fv \quad \Gamma\Sigma\neg Gvz \to Fw\ (S_2)}{\Gamma\Sigma,\ \neg Gvz \to Fv\&Fw}}{\Gamma\Sigma \to \neg Gvz \supset Fv\&Fw}}{\Gamma\Sigma \to \exists z(\neg Gvz \supset Fv\&Fw)}$$

$$\Downarrow$$

$$\dfrac{\dfrac{\Gamma \to \exists z(\neg Gvz \supset Fv\&Fw)}{\Gamma \to \forall v\exists z(\neg Gvz \supset Fv\&Fw)}\ \pi_\forall}{\Gamma \to \forall w\forall v\exists z(\neg Gvz \supset Fv\&Fw)}$$

In this connection it can be assumed (in view of the reversibility of the rule $\exists \to$) that Σ consists of formulas of the form

$$\forall u\exists x(\neg(Gux\&\beta = u) \supset F\beta), \tag{I}$$

$$\neg(Gax\&\beta = \alpha) \supset F\beta. \tag{II}$$

In view of the weededness of the proof D, the succedent formulas Fv and Fw in the sequents (S_1) and (S_2) are traceable to axioms. This implies that in the sequents located above (S_1) only those formulas of types I and II in which β is v are traceable to axioms. It is sufficient to consider only formulas of type II. Let us consider the axiom corresponding to the formula F (the F-axiom). It has the form $F_\gamma \to F_\gamma$, where the antecedent F_γ originates from F_β and the succedent one from Fv (the latter in view of the absence above $\pi_\forall$ of rules introducing new positive occurrences of F). If β is different from v, an equality containing v must be encountered above (S_1) in the antecedent. Such equalities neither occur in Γ (because of the restriction on variables in $\pi_\forall$) nor occur as members of the list Σ, and cannot appear in the section from (S_1) to an axiom since Fv would be annihilated with that (in at least one of the premises). Therefore β is v. Analogously β is w in all formulas of types (I) and (II) which are traceable to axioms and located in (S_2) and higher. Let us note in addition that only one of the formulas of type (II) over each of (S_1) and (S_2) is traceable to axioms: the lowest $(\supset \to)$-inference "annihilates" the succedent F-formula in the left premise.

In view of the restriction on variables in the rule $(\exists \to)$, the variable (substituted for) x in a formula of type II which is traceable to an axiom over (S_1) is different from the variable (substituted for) x in the corresponding formula over (S_2).

Therefore one of them (for definiteness, the one which is over (S_1)) is different from z. The corresponding G-axiom must be of the form $G\alpha'x' \to G\alpha'x'$, where the antecedent formula originates from $G\alpha x$ and the succedent one from Gvz. The only equality which can "help" the appearance of this axiom is $v = \alpha$. (Let us recall that β is v.) Since because of the restriction on variables x is different from v and α, the only possibility, v is z (under substitution of z for α), is insufficient.

Appendix

On removing structural rules from Gentzen calculi

Let us consider a one-succedent variant J of the intuitionistic predicate calculus which does not contain structural rules (for example, the calculus LJ_1 of [4] or the calculus K_1^* of [3]), and let us prove its equivalence to the calculus G. These calculi differ from the calculus G basically in the formulation of the rules $(\forall \to)$, $(\supset \to)$, $(\neg \to)$ and $(\& \to)$:

$$\frac{\Gamma,\ \forall xA,\ [A]^x_t,\ \Sigma \to \Delta}{\Gamma,\ \forall xA,\ \Sigma \to \Delta},\quad \frac{\Gamma,\ A,\ B,\ \Sigma \to \Delta}{\Gamma,\ A\&B,\ \Sigma \to \Delta},$$

$$\frac{\Gamma,\ (A \supset B),\ \Sigma \to A \qquad \Gamma,\ B,\ \Sigma \to \Delta}{\Gamma,\ (A \supset B),\ \Sigma \to \Delta},\quad \frac{\Gamma,\ \neg A,\ \Sigma \to A}{\Gamma,\ \neg A,\ \Sigma \to \Delta}.$$

Another difference consists in the absence of structural rules. Let us note that thinning and permutation rules are simulated in J without altering the proof structure: thinning in an antecedent is simulated by multiplying (§1) by a corresponding purely antecedent sequent. To simulate the rule $(\to Y)$ the operation of multiplying by a purely succedent sequent must be introduced: the "multiplier" is written into all empty succedents below which there are no nonempty succedents in the original inference.

Permutation in the bottom sequent is simulated by corresponding permutations in higher placed sequents. Therefore, only cancellation of repetitions presents a difficulty, and it is sufficient to establish the following assertion.

THEOREM A.1. *It is possible to construct with respect to every proof in the calculus G a proof of the same sequent in the calculus* $J + (\to Y,\ Y\to,\ \Pi\to)$.

Indeed, let us consider a proof D satisfying Condition (A) (cf. §1). Let us carry out cancellation of repetitions in D, i.e. let us replace sequents of the form $\Gamma C\Gamma_1 C\Gamma_2 \cdots C\Gamma_n \to \Delta$ by $\Gamma C\Gamma_1\Gamma_2 \cdots \Gamma_n \to \Delta$. What happens to the rule applications? Succedent inferences either turn into inferences by the same rules or may be filled out to such with the help of thinning. Also preserved are the antecedent inferences whose principal formulas were not repeated in the conclusion. (Cancelling repetitions in the premises could spoil only semidegenerate inferences.) It remains to consider the case

$$L\,\frac{E_1A\Gamma_1 \to \Delta_1\ (E_2A\Gamma_2 \to \Delta_2)}{AA\Gamma \to \Delta},\quad L'\,\frac{E_1A\Gamma_1' \to \Delta_1\ \ (E_2A\Gamma_2' \to \Delta_2)}{A\Gamma' \to \Delta}.$$

In view of the weededness of the proof, A cannot be introduced into all of the premises of L by thinning. Therefore $L \neq \forall \rightarrow, \exists \rightarrow$; since A is traceable to axioms, there would in this case be a degenerate inference (with principal formula A by the same rule as L) above L. If $L = \neg\rightarrow, \vee \rightarrow$, then L' is an inference by the rule of the same name in the J calculus. If L is an $(\supset \rightarrow)$-inference, A is introduced into the right premise by a thinning. (Otherwise L would be degenerate.) Eliminating A from the right premise, we obtain an inference by the rule $\supset \rightarrow$ of the calculus J. Finally, if $L = (\& \rightarrow)$, after replacing all preimages of the occurrence of A under consideration in the premise by the "missing" side formula, i.e. by the formula C if the frame of L is of the form $B\&C \rightarrow \vdash B \rightarrow$, we again obtain a proof: L' turns into $\& \rightarrow$, the axioms are not spoiled since A does not occur in them (by virtue of Condition (A), only elementary formulas occur in them), and inferences with principal formula A turn into repetitions of the sequents

$$\frac{C,\ \Gamma \rightarrow \Delta}{B\&C,\ \Gamma \rightarrow \Delta}, \quad \frac{C,\ \Gamma \rightarrow \Delta}{C,\ \Gamma \rightarrow \Delta},$$

since inferences with the side formula B would be degenerate.

After adding thinnings (in order to reconstruct the rules $\vee \rightarrow, \neg\rightarrow, \supset \rightarrow$ and $\& \rightarrow$ of the calculus G into corresponding rules of J), we obtain an inference in the required calculus.

Let us note that it is possible by the same method (within the limits of the general scheme) to give a new proof of the removability of structural rules (in the first place, of the cancellation of repetitions) from multi-succedent variants of the classical, the intuitionistic, and the minimal predicate calculi. Corresponding modifications of the concept of degenerate inference and of the operation of multiplying a proof by a sequent can be constructed without particular difficulty.

BIBLIOGRAPHY

1. G. E. Minc, *Skolem's method of elimination of positive quantifiers in sequential calculi,* Dokl. Akad. Nauk SSSR **169** (1966), 24–27 = Soviet Math. Dokl. **7** (1966), 861–864. MR **34** #5648.

2. S. C. Kleene, *Permutability of inferences in Gentzen's calculi LK and LJ. Two papers on the predicate calculus,* Mem. Amer. Math. Soc. No. 10 (1952); Russian transl. in Mathematical Theory of Logical Inference, "Nauka", Moscow, 1967, pp. 208–236. MR **14**, 439; **36** #4957.

3. V. P. Orevkov, *On Glivenko sequent classes,* Trudy Mat. Inst. Steklov. **98** (1968), 131–154 = Proc. Steklov Inst. Math. **98** (1968), 147–176. MR **39** #2606; **43** #4620.

4. H. B. Curry, *Foundations of mathematical logic,* McGraw-Hill, New York, 1963; Russian transl., Moscow, 1969. MR **26** #6036.

5. G. E. Minc, *Herbrand's theorem,* Math. Theory of Logical Deduction, "Nauka", Moscow, 1967, pp. 311-350. (Russian) MR **36** #4958.

6. G. E. Minc and V. P. Orevkov, *An extension of the theorems of Glivenko and Kreisel to a certain class of formulas of predicate calculus,* Dokl. Akad. Nauk SSSR **152** (1963), 553–554 = Soviet Math. Dokl. **4** (1963), 1365–1366. MR **27** #4737.

7. S. A. Kripke, *Semantical analysis of intuitionistic logic.* I, Formal Systems and Recursive Functions (Proc. Eighth Logic Colloq., Oxford, 1963), North-Holland, Amsterdam, 1965, pp. 92–130. MR **34** #1184.

Translated by F. M. GOLDWARE

Trudy Mat. Inst. Steklov.
121 (1972)

Proc. Steklov Inst. Math.
121 (1972)

UNDECIDABLE CLASSES OF FORMULAS FOR THE CONSTRUCTIVE PREDICATE CALCULUS. I

UDC 51.01: 164

V. P. OREVKOV

ABSTRACT. A sequent is called absolute if there is not more than one formula in its succedent and it does not contain any occurrences of the symbols $\supset$ or $\neg$ or any positive occurrences of the quantifier $\forall$. It is shown that an absolute sequent is derivable in the classical predicate calculus if and only if it is derivable in the constructive (or in the minimal) predicate calculus. Using methods due to Gentzen, we prove that the class of absolute sequents of the form

$$A \to \exists x_1 \ldots x_n (P_1 (x_2, \ldots, x_n) \& P_2 (x_1, \ldots, x_n)),$$

where the formula A does not contain any predicate variables other than P_1 or P_2, is a reduction class for the classical, as well as for the constructive, predicate calculus.

Bibliography: 12 items.

CONTENTS. Introduction. § 1. Logical calculi and axiomatic theories. § 2. Auxiliary definitions and lemmas. § 3. Absolute sequents. Bibliography.

Introduction

It is well known that the derivability problem is algorithmically decidable in the classical predicate calculus without functional symbols, for the class of all formulas containing only unary predicate variables. The main aim of this paper is to give a constructive proof of the undecidability of the derivability problem in the constructive predicate calculus without functional symbols, for the class of all formulas containing only a single unary (and no other) predicate variable.

Most of the results of this paper were stated in [1] and [2]. The main results were presented in a talk at the International Congress of Mathematicians in 1966, and published in [3].

All unexplained special terms and notation are to be understood as in [4]. In each section, theorems and lemmas are designated by a pair of natural numbers, the first of which coincides with the number of the section, and the second of which is the ordinal number of the given theorem or lemma within the section.

§ 1. Logical calculi and axiomatic theories

We shall call (recursively) enumerable sets of formulas of the predicate calculus,

---

AMS (MOS) subject classifications (1970). Primary 02C15, 02G05; Secondary 02B10.

without functional symbols or equality, *logical calculi*. We shall say that a formula is derivable in a logical calculus if it belongs to the given enumerable set. We shall say that a sequent is derivable in a logical calculus if its formula image (see [5], §3, or [6], Chapter V, §1) is derivable there.

The main examples of logical calculi which will play a role in this paper are the classical, constructive, minimal and positive predicate calculi, without functional symbols. The positive (called absolute by some authors) predicate calculus is obtained from the constructive predicate calculus by deleting the symbol $\neg$ from the language, and deleting those axioms and rules of inference in which this symbol occurs from the list of axioms and rules. It is clear that a formula in which the symbol $\neg$ does not occur is derivable in the constructive predicate calculus if and only if it is derivable in the positive predicate calculus. In what follows, the letters **C**, **K**, **M** and **A** will denote the classical, constructive, minimal and positive predicate calculi, respectively. We shall call the calculus obtained from **K** by deleting those axioms and rules of inference in which the symbol $\neg$ occurs the *formal $\neg$-extension* of **A**, and shall denote it by $\mathbf{A}^{\neg}$.

We shall say that the logical calculus $\mathfrak{K}$ is an *extension* of the calculus $\mathfrak{R}$ if every formula derivable in $\mathfrak{R}$ is also derivable in $\mathfrak{K}$. We shall say that the logical calculus $\mathfrak{P}$ is *situated between* the calculi $\mathfrak{K}$ and $\mathfrak{R}$ if $\mathfrak{P}$ is an extension of $\mathfrak{K}$ while $\mathfrak{R}$ is an extension $\mathfrak{P}$. For example, the calculi **K** and **M** are situated between the calculi $\mathbf{A}^{\neg}$ and **C**.

Let us carry over the terminology used in the theory of decidable fragments of the classical predicate calculus to the case of an arbitrary logical calculus.

Let $\mathfrak{N}$ be a class of formulas or sequents of a predicate calculus without functional symbols, and let $\mathfrak{P}$ be a logical calculus. We shall call $\mathfrak{N}$ a *decidable class* for the calculus $\mathfrak{P}$ if an algorithm recognizing which elements of $\mathfrak{N}$ are derivable in $\mathfrak{P}$ is possible. Otherwise $\mathfrak{N}$ will be called an *undecidable class* for $\mathfrak{P}$. We shall call $\mathfrak{N}$ a *reduction class* for $\mathfrak{P}$ if it is possible to construct an algorithm φ such that, for any formula or sequent S, φ transforms S into an element of $\mathfrak{N}$, and $\varphi(S)$ is derivable in $\mathfrak{P}$ if and only if S is derivable in $\mathfrak{P}$. We note that if the class of all formulas is an undecidable class for $\mathfrak{P}$, then so is every reduction class for $\mathfrak{P}$.

Let $\mathfrak{K}$ and $\mathfrak{R}$ be logical calculi, and let $\mathfrak{N}$ be a class of sequents of a predicate calculus without functional symbols. We shall say that $\mathfrak{N}$ is a *Glivenko class* for the calculi $\mathfrak{K}$ and $\mathfrak{R}$ if a sequent of $\mathfrak{N}$ is derivable in $\mathfrak{K}$ if and only if it is derivable in $\mathfrak{R}$. We shall call a logical calculus *creative* if it is an m-universal (creative) (cf., for example, [7], §§8.1, 8.2) enumerable set of derivable formulas.

Let us state sufficient conditions for a class of sequents to be a reduction class. It is not difficult to prove the following assertions.

LEMMA 1.1. *If the calculus* $\mathfrak{R}$ *is creative and the class* $\mathfrak{M}$ *is a Glivenko class for* $\mathfrak{K}$ *and* $\mathfrak{R}$ *and a reduction class for* $\mathfrak{R}$, *then* $\mathfrak{M}$ *is a reduction class for any logical calculus situated between* $\mathfrak{K}$ *and* $\mathfrak{R}$.

LEMMA 1.2. $\mathfrak{M}$ *is a reduction class for every logical calculus situated between* $\mathfrak{K}$ *and* $\mathfrak{R}$, *if* $\mathfrak{R}$ *is creative and one can find a class of sequents* $\mathfrak{N}$ *and an algorithm* φ *such that the following conditions hold*:

1) $\mathfrak{N}$ *is a reduction class for* $\mathfrak{R}$ *and a Glivenko class for* $\mathfrak{K}$ *and* $\mathfrak{R}$.

2) φ *transforms each sequent of* $\mathfrak{N}$ *into a sequent of* $\mathfrak{M}$.

3) *given any sequent* S *of* $\mathfrak{N}$, *if* S *is derivable in* $\mathfrak{K}$, *then* $\varphi(S)$ *is derivable in* $\mathfrak{K}$, *and if* $\varphi(S)$ *is derivable in* $\mathfrak{N}$, *then* S *is derivable in* $\mathfrak{R}$.

We shall call a finite list of predicate symbols a signature. Let σ be a fixed signature. We shall call formulas (respectively, sequents) of a predicate calculus, without functional symbols or equality, which contain no predicate symbols distinct from those in σ, *formulas* (*sequents*) *of signature* σ.

Let us consider a fixed logical calculus $\mathfrak{P}$ and a fixed constructive set $\mathfrak{A}$ of formulas of signature σ. We shall say that *the sequent* S *of signature* σ *is derivable in the* $\mathfrak{P}$-*theory of signature* σ *with axiom system* $\mathfrak{A}$ (or, more briefly, in the $\mathfrak{P}$-theory $[\mathfrak{A}]_\sigma$) if one can find a list (possibly empty) of formulas of $\mathfrak{A}$ such that prefixing this list to the antecedent of S yields a sequent derivable in $\mathfrak{P}$. We shall call a class of sequents of signature σ *decidable* with respect to the $\mathfrak{P}$-theory $[\mathfrak{A}]_\sigma$ if an algorithm recognizing which sequents of this class are derivable in the $\mathfrak{P}$-theory $[\mathfrak{A}]_\sigma$ is possible. Otherwise we shall call it *undecidable* with respect to $[\mathfrak{A}]_\sigma$. We shall say that the $\mathfrak{P}$-theory $[\mathfrak{A}]_\sigma$ is *decidable* if the class of all sequents of signature σ is decidable with respect to it. Otherwise, we shall say that it is *undecidable.*

We call a logical calculus $\mathfrak{P}$ *proper* if the following conditions hold:

1) For any formula A of the predicate calculus without functional symbols or equality, the sequent $A \to A$ is derivable in $\mathfrak{P}$.

2) The cut rule is allowable([1]) in $\mathfrak{P}$.

3) The antecedent structural rules of thinning, contraction and interchange are allowable in $\mathfrak{P}$.

Let $\mathfrak{A}$ and $\mathfrak{B}$ be two constructive sets of formulas of signature σ. We shall say that the $\mathfrak{P}$-theory $[\mathfrak{A}]_\sigma$ is a *subtheory* of the $\mathfrak{P}$-theory $[\mathfrak{B}]_\sigma$ if every sequent derivable in the former is also derivable in the latter. In [8] a classical theory is called hereditarily undecidable if it, as well as each of its subtheories with the same

---

([1]) A rule is called allowable in the calculus $\mathfrak{P}$ if for any application of the rule the derivability in $\mathfrak{P}$ of all the premises of this application implies the derivability in $\mathfrak{P}$ of the conclusion of this application.

signature, is undecidable. There it is proved that an undecidable and finitely axiomatizable classical theory is hereditarily undecidable.

This theorem may be carried over to the case of an arbitrary logical calculus in the form of the following assertion.

THEOREM 1.1. *For any signature σ, finite list $\mathfrak{S}$ of formulas of σ, constructive set $\mathfrak{A}$ of formulas of σ and logical calculus $\mathfrak{P}$, if the $\mathfrak{P}$-theory $[\mathfrak{A}]_\sigma$ is a subtheory of the $\mathfrak{P}$-theory $[\mathfrak{S}]_\sigma$, the $\mathfrak{P}$-theory $[\mathfrak{S}]_\sigma$ is undecidable and the calculus $\mathfrak{P}$ is proper, then the $\mathfrak{P}$-theory $[\mathfrak{A}]_\sigma$ is undecidable.*

PROOF. Let the signature σ, the finite list $\mathfrak{S}$ of formulas of σ, the set $\mathfrak{A}$ of formulas of σ and the logical calculus $\mathfrak{P}$ satisfy the conditions of the theorem. Let S be a sequent of σ. Let S' denote the sequent obtained from S by prefixing the list $\mathfrak{S}$ to its antecedent. We shall show that S is derivable in the $\mathfrak{P}$-theory $[\mathfrak{S}]_\sigma$ if and only if S' is derivable in $[\mathfrak{A}]_\sigma$. It is obvious that this implies Theorem 1.1. Assume that S is derivable in $[\mathfrak{S}]_\sigma$. It then follows from the allowability of the antecedent structural rules that S' is derivable in $\mathfrak{P}$ and therefore in $[\mathfrak{A}]_\sigma$.

Assume that the sequent S' is derivable in the $\mathfrak{P}$-theory $[\mathfrak{A}]_\sigma$. One can then find a list Π of formulas of $\mathfrak{A}$, which, when prefixed to the antecedent of S', yields a sequent derivable in $\mathfrak{P}$. If Π is empty, then S' is derivable in $\mathfrak{P}$ and therefore S is derivable in $[\mathfrak{S}]_\sigma$. Let us suppose that Π is not empty, and that it is written out in the form $A_1, \cdots, A_n$.

It follows from condition 1) of the definition of a proper logical calculus that the sequents $\to A_1, \cdots, \to A_n$ are derivable in $[\mathfrak{A}]_\sigma$. Since the latter is a subtheory of $[\mathfrak{S}]_\sigma$, one can construct lists $\Gamma_1, \cdots, \Gamma_n$ of formulas of $\mathfrak{S}$ such that the sequents $\Gamma_1 \to A_1, \cdots, \Gamma_n \to A_n$ are derivable in $\mathfrak{P}$. It remains to make use of the allowability of the cut rule and the antecedent structural rules in $\mathfrak{P}$. Theorem 1.1 is proved.

Let $\mathfrak{R}$ be a class of sequents of signature σ. We shall call $\mathfrak{R}$ a *reduction class* with respect to the $\mathfrak{P}$-theory $[\mathfrak{A}]_\sigma$ if we can construct an algorithm φ transforming each sequent into an element of $\mathfrak{R}$, such that for every sequent S, $\varphi(S)$ is derivable in the $\mathfrak{P}$-theory $[\mathfrak{A}]_\sigma$ if and only if S is derivable in $\mathfrak{P}$. Note that in case $\mathfrak{A}$ is the empty set, a decidable (respectively, reduction) class with respect to $[\mathfrak{A}]_\sigma$ will be a decidable (reduction) class for $\mathfrak{P}$. It is obvious that any class of sequents of signature σ containing some reduction class with respect to $[\mathfrak{A}]_\sigma$ will also be a reduction class with respect to $[\mathfrak{A}]_\sigma$.

We shall call $\mathfrak{R}$ an *hereditary reduction class* with respect to $[\mathfrak{A}]_\sigma$ if it is possible to construct an algorithm φ such that for any sequent S the following conditions hold:

1) φ transforms S into a sequent of $\mathfrak{R}$.

2) If S is derivable in $\mathfrak{P}$, so is $\varphi(S)$.

3) If $\varphi(S)$ is derivable in $[\mathfrak{A}]_\sigma$, then S is derivable in $\mathfrak{P}$.

It is obvious that every hereditary reduction class with respect to a $\mathfrak{P}$-theory $[\mathfrak{A}]_\sigma$ will be a reduction class with respect to each of its subtheories and, therefore, a reduction class for the calculus $\mathfrak{P}$. In what follows, we shall often make use of the following assertion.

LEMMA 1.3. *For any signatures σ and τ, set $\mathfrak{A}$ of formulas of signature σ, set $\mathfrak{B}$ of formulas of signature τ, logical calculus $\mathfrak{P}$, class $\mathfrak{M}$ of sequents of σ and creative logical calculus $\mathfrak{R}$, which is an extension of $\mathfrak{P}$, $\mathfrak{M}$ is an hereditary reduction class with respect to the $\mathfrak{P}$-theory $[\mathfrak{A}]_\sigma$ if one can find a class $\mathfrak{N}$ of sequents of τ and an algorithm φ such that the following conditions hold*:

1) *$\mathfrak{N}$ is a Glivenko class for the calculi $\mathfrak{P}$ and $\mathfrak{R}$.*

2) *$\mathfrak{N}$ is an hereditary reduction class with respect to the $\mathfrak{R}$-theory $[\mathfrak{B}]_\tau$.*

3) *φ transforms each sequent of $\mathfrak{N}$ into a sequent of $\mathfrak{M}$.*

4) *Given any sequent S of $\mathfrak{N}$, if S is derivable in $\mathfrak{P}$, then $\varphi(S)$ is derivable in $\mathfrak{P}$, and if $\varphi(S)$ is derivable in the $\mathfrak{R}$-theory $[\mathfrak{A}]_\sigma$, then S is derivable in the $\mathfrak{R}$-theory $[\mathfrak{B}]_\tau$.*

This lemma follows directly from the appropriate definitions.

§2. Auxiliary definitions and lemmas

Consider an arbitrary realization of the rule $\to\forall$ of the calculus $\mathbf{C}^*$ (see [4], §1)

$$\frac{\Gamma\to\Delta',\ [A]^x_a,\ \Delta''}{\Gamma\to\Delta',\ \forall xA,\ \Delta''}\to\forall.$$

We shall call the variable a, explicitly indicated in the premise of this realization, the *chacteristic variable of the realization.* The relation "chacteristic variable of the realization" is defined analogously for realizations of the rule $\exists\to$ of the calculus $\mathbf{C}^*$ and for realizations of the rules $\to\forall$ and $\exists\to$ of the calculi $\mathbf{C}^+$, $\mathbf{K}_1^*$ and $\mathbf{K}_1^+$. The following assertion is obvious.

LEMMA 2.1. *If a deduction in any one of the calculi $\mathbf{C}^*$, $\mathbf{C}^+$, $\mathbf{K}_1^*$ or $\mathbf{K}_1^+$ has the pure variable property* ([9], §78), *then the characteristic variables of distinct applications of the rules $\to\forall$ and $\exists\to$ in this deduction are distinct.*

Let X and Y be lists of variables (possibly with repetitions). We shall say that X *is contained in* Y, and write $X\subseteq Y$, if every variable in X is also in Y. We shall say that X and Y are *equivalent* if X is contained in Y and Y is contained in X.

Let D be a tree form deduction in any one of the calculi $\mathbf{C}^*$, $\mathbf{C}^+$, $\mathbf{K}_1^*$ or $\mathbf{K}_1^+$, and let α and β be variables. We shall say that α *is higher than* β *in* D, and write $\beta \prec_D \alpha$, if β occurs free in some sequent S of D, while there is no free occurrence of α in any sequent which is not higher than S in D. It is not difficult to prove the following lemma.

LEMMA 2.2. *For every deduction D in any one of the calculi* $\mathbf{C}^*$, $\mathbf{C}^+$, $\mathbf{K}_1^*$ *or* $\mathbf{K}_1^+$, *the relation* $\prec_D$ *is irreflexive and transitive.*

In the course of this paper, only tree form deductions will be constructed in the calculi $\mathbf{C}^*$, $\mathbf{C}^+$, $\mathbf{K}_1^*$ and $\mathbf{K}_1^+$. An occurrence of a formula in a sequent will only mean an occurrence of a formula as a member of a sequent. An occurrence of a subformula in a sequent will mean either an occurrence of a formula as a member of a sequent or an occurrence of a formula as a subformula of a formula occurring in a sequent.

We shall call a sequent *positive* if it does not contain any occurrences of the symbols $\supset$ or $\neg$. We shall call the sequent S *quasi-positive* if S does not contain any occurrences of the symbol $\supset$, and there are no occurrences of the symbols &, $\vee$, $\forall$ or $\exists$ in the scope of any occurrence of the symbol $\neg$ in S. We shall call a positive (quasi-positive) sequent an *absolute* (*quasi-absolute*) sequent if it does not contain any positive occurrences of the quantifier $\forall$, and is a singulary sequent, i.e. if there is not more than one formula in its succedent.

Let S be an arbitrary sequent of the predicate calculus, let $\mathfrak{C}_0, \mathfrak{C}_1, \cdots, \mathfrak{C}_k$ be a regular * partition for the sequent S and let D be a deduction of S in any of the calculi $\mathbf{C}^*$, $\mathbf{C}^+$, $\mathbf{K}_1^*$ or $\mathbf{K}_1^+$. We shall call an application of a rule in D an *i-midapplication* $(0 \leqslant i \leqslant k)$ for the partition $\mathfrak{C}_0, \mathfrak{C}_1, \cdots, \mathfrak{C}_k$ if it belongs to some occurrence from the classes $\mathfrak{C}_i, \cdots, \mathfrak{C}_k$, but there are no applications of any rules belonging to occurrences from the classes $\mathfrak{C}_i, \cdots, \mathfrak{C}_k$ higher than any of its premises. Such premises will be called *i-midsequents* for the partition $\mathfrak{C}_0, \mathfrak{C}_1, \cdots, \mathfrak{C}_k$. In what follows, k-midsequents and k-midapplications of rules will be called *midsequents* and *midapplications* of rules.

§3. Absolute sequents

Note the following important property of absolute sequents.

THEOREM 3.1. *Any absolute sequent derivable in* $\mathbf{C}$ *is also derivable in* $\mathbf{A}$.

PROOF. Let S be an absolute sequent. Then S belongs to the σ-class $\{\supset^+, \neg^+, \forall^+\}$. According to Theorem 4.1 of (4), this is a completely Glivenko class. Therefore from the derivability of S in $\mathbf{C}$ follows its derivability in $\mathbf{K}$. To complete the proof of Theorem 3.1, note that all rules and axioms (except for the rules $\neg \rightarrow$ and $\rightarrow \neg$) coincide for the Gentzen variants of the calculi $\mathbf{K}$ and $\mathbf{A}$ (see, for example, [10]) in their usual formulations.

Theorem 3.1 and the fact that $\mathbf{C}$, $\mathbf{K}$ and $\mathbf{M}$ are extensions of $\mathbf{A}$ imply the following assertion.

*Editor's note. The Russian term is правильное, which is translated as "correct" in the English version of [4] (specifically §3).

COROLLARY. *The class of absolute sequents is a Glivenko class for the calculi* $\mathfrak{P}$ *and* $\mathfrak{R}$, *where* $\mathfrak{P}$ *and* $\mathfrak{R}$ *are any of the calculi* **C**, **K**, **M** *or* **A**.

We shall show that the class of absolute sequents is a reduction class for any logical calculus situated between **A** and **C**.

THEOREM 3.2. *The class of absolute sequents of the form*

$$QM \rightarrow \exists x_1 \ldots x_n (P_1(x_1, \ldots, x_n) \& P_2(x_1, \ldots, x_n)), \tag{1}$$

where n *is a positive integer,* $x_1, \cdots, x_n$ *are mutually distinct object variables,* P_1 *and* P_2 *are distinct n-ary predicate variables,* Q *is a prefix* ([11], §39) *and* M *is a disjunction of conjunctions of elementary formulas containing no predicate variables other than* P_1 *or* P_2, *is a reduction class for any logical calculus situated between* **A** *and* **C**.

PROOF. According to Lemma 1.1 and the Corollary to Theorem 3.1, it is sufficient to ascertain that the class of sequents under consideration is a reduction class for **C**. Let us prove this.

Indeed, let A be an arbitrary formula of the predicate calculus. Construct a prefix Q and a disjunction of elementary conjunctions N such that A is equivalent to $\neg QN$ in **C**. Write out the list of all predicate variables occurring in $\neg QN$ as follows: $R_1, \cdots R_s$. Denote the rank of the predicate variable R_i $(1 \leqslant i \leqslant s)$ by m_i, and denote the number $\max(m_1, \cdots, m_s) + 1$ by n. Introduce two new distinct n-ary predicate variables P_1, P_2 and s new mutually distinct object variables $a_1, \cdots, a_s$. Construct the sequent

$$QN' \rightarrow \exists x_1 \ldots x_n (P_1(x_1, \ldots, x_n) \& P_2(x_1, \ldots, x_n)), \tag{2}$$

where $x_1, \cdots, x_n$ are mutually distinct object variables, and N' is obtained from N by replacing each occurrence of a formula of the form $\neg R_i(\alpha_1, \cdots, \alpha_{m_i})$ by the formula

$$P_2(\alpha_1, \ldots, \alpha_{m_i}, \underbrace{a_i, \ldots, a_i}_{n-m_i \text{ times}}), \tag{3}$$

and then replacing each occurrence of a formula of the form $R_i(\alpha_1, \cdots, \alpha_{m_i})$ by the formula

$$P_1(\alpha_1, \ldots, \alpha_{m_i}, \underbrace{a_i, \ldots, a_i}_{n-m_i \text{ times}}). \tag{4}$$

We shall prove that the sequent (2) is derivable in **C** if and only if the sequent $QN \rightarrow$ (and therefore the formula A) is derivable in **C**.

Let the sequent (2) be derivable in **C**, and let D be a proof of (2) in **C***. In each sequent of D, replace every occurrence of a subformula of the form

$P_2(\alpha_1, \cdots, \alpha_n)$ by the formula $\neg P_1(\alpha_1, \cdots, \alpha_n)$, obtaining a proof of the sequent

$$QN'' \rightarrow \exists x_1 \ldots x_n (P_1(x_1, \ldots, x_n) \& \neg P_1(x_1, \ldots, x_n)).$$

Note that the sequent

$$\exists x_1 \ldots x_n (P_1(x_1, \ldots, x_n) \& \neg P_1(x_1, \ldots, x_n)) \rightarrow$$

is derivable in **C**.

Hence it follows from the allowability of the cut rule in **C** that the sequent $QN'' \rightarrow$ is derivable in **C**. Denote a proof of $QN'' \rightarrow$ in $\mathbf{C}^*$ by D'. It is obvious that all elementary subformulas occurring in sequents of D' have the form (4). In each sequent of D', replace every occurrence of a subformula of the form $P_1(\alpha_1, \cdots, \alpha_n)$ by the formula $R_i(\alpha_1, \cdots, \alpha_{m_i})$ for all i $(1 \leqslant i \leqslant s)$, obtaining a proof of the sequent $QN \rightarrow$.

Let $QN \rightarrow$ be derivable in **C**. Partition the occurrences of logical symbols in QN into two classes: put the occurrences of the sign $\neg$ into the first class, and the occurrences of the other logical connectives into the second class. It is clear that this gives us a classically regular partition for the sequent $QN \rightarrow$. Let D'' be a proof of $QN \rightarrow$ in $\mathbf{C}^*$ which is compatible with the partition described above. Delete from D'' those parts which are higher than midsequents. After that, in each sequent of the configuration thus obtained, replace all occurrences of subformulas of the form $\neg R_i(\alpha_1, \cdots, \alpha_{m_i})$ by formulas of the form (3), and then replace all occurrences of subformulas of the form $R_i(\alpha_1, \cdots, \alpha_{m_i})$ by formulas of the form (4). After that, write down the formula

$$\exists x_1 \ldots x_n (P_1(x_1, \ldots, x_n) \& P_2(x_1, \ldots, x_n))$$

in the succedents (they are empty) of all sequents of the configuration thus obtained.

Denote the resulting configuration by $\mathfrak{D}$. We have transformed all midsequents into sequents of the form (up to permutations)

$$\Gamma_1, P_1(\alpha_1, \ldots, \alpha_n), \Gamma_2, P_2(\alpha_1, \ldots, \alpha_n), \Gamma_3 \rightarrow \exists x_1 \ldots x_n (P_1(x_1, \ldots, x_n) \& P_2(x_1 \ldots, x_n)).$$

It is obvious that sequents of this form are derivable in **C**. Writing proofs in $\mathbf{C}^*$ of these sequents above $\mathfrak{D}$, we obtain a proof of the sequent (2). Theorem 3.2 is proved.

REMARK. One can prove that a formula A is refutable on a finite model if and only if the sequent of the form (1) corresponding to A in the manner indicated above is also refutable on a finite model. Hence, making use of results from (12), we obtain the following assertion:

The class of absolute sequents of the form (1) *derivable in* A *is effectively inseparable from the class of absolute sequents of the form* (1) *refutable on finite models.*

BIBLIOGRAPHY

1. V. P. Orevkov, *Certain reduction classes and solvable classes of sequents for the constructive predicate calculus,* Dokl. Akad. Nauk SSSR **163** (1965), 30–32 = Soviet Math. Dokl. **6** (1965), 888–890. MR **33** #51.

2. S. Ju. Maslov, G. E. Minc and V. P. Orevkov, *Unsolvability in the constructive predicate calculus of certain classes of formulas containing only monadic predicate variables,* Dokl. Akad. Nauk SSSR **163** (1965), 295–297 = Soviet Math. Dokl. **6** (1965), 918–920. MR **33** #52.

3. V. P. Orevkov, *Some reducible classes for the constructive and minimal predicate calculi,* Abstracts of Brief Scientific Communications, Internat. Congress Math., Moscow, 1966, Section I, pp. 22–23. (Russian)

4. ———, *On Glivenko sequent classes,* Trudy Mat. Inst. Steklov. **98** (1968), 131–154 = Proc. Steklov Inst. Math. **98** (1968), 147–176. MR **39** #2606; **43** #4620.

5. G. E. Minc, *Herbrand's theorem,* Math. Theory of Logical Deduction, "Nauka", Moscow, 1967, pp. 311–350. (Russian) MR **36** #4958.

6. G. Gentzen, *Untersuchungen über das logische Schliessen,* Math. Z. **39** (1934/35), 176–210, 405–431; Russian transl. in Mathematical Theory of Logical Inference, "Nauka", Moscow, 1967, pp. 9–74; English transl. in M. E. Szabo, *The collected papers of Gerhard Gentzen,* Studies in Logic and the Foundations of Math., North-Holland, Amsterdam, 1969, pp. 68–103. MR **36** #4957; **41** #6660.

7. A. I. Mal'cev, *Algorithms and recursive functions,* "Nauka", Moscow, 1965; English transl., Wolters-Noordhoff, Groningen, 1970. MR **34** #2453.

8. A. Tarski, A. Mostowski and R. M. Robinson, *Undecidable theories,* Studies in Logic and the Foundations of Math., North-Holland, Amsterdam, 1953. MR **15**, 384.

9. S. C. Kleene, *Introduction to metamathematics,* Van Nostrand, Princeton, N. J., 1952; Russian transl., IL, Moscow, 1957. MR **14**, 525; **19**, 2.

10. H. B. Curry, *Foundations of mathematical logic,* McGraw-Hill, New York, 1963; Russian transl., Moscow, 1969. MR **26** #6036.

11. A. Church, *Introduction to mathematical logic.* Vol. I, Princeton Univ. Press, Princeton, N. J., 1956; Russian transl., Moscow, 1960. MR **18**, 631.

12. B. A. Trahtenbrot, *On recursive separability,* Dokl. Akad. Nauk SSSR **88** (1953), 953–956. (Russian) MR **16**, 436.

Translated by M. GREENDLINGER

Trudy Mat. Inst. Steklov.
121 (1972)

Proc. Steklov Inst. Math.
121 (1972)

A SEQUENTIAL VARIANT OF R. M. ROBINSON'S ARITHMETIC SYSTEM NOT CONTAINING CUT RULES

A. Ja. PLJUŠKEVIČENE

UDC 51.01:164

ABSTRACT. A sequential variant of Raphael M. Robinson's arithmetic system is constructed which does not contain cut rules, including cut rules introducing specific axioms of the theory (we have in mind that the axioms of the theory are not written in the antecedent of the formula being proved). The proof of the equivalence of the constructed calculus with Robinson's arithmetic system is such that it implies a new proof of the consistency of Robinson's system. As an example of an intermediate calculus, we construct a sequential variant of Robinson's arithmetic system containing only one cut rule (corresponding to the specific axiom $\forall x(x = 0 \vee \exists y(x = y')))$. A number of theorems on the specialization of the form of a deduction in the intermediate calculus are proved.

Bibliography: 8 items.

Introduction

Of primary significance in the solution of the deduction problem in various axiomatically given mathematical theories is the problem of dealing with specific axioms of this theory. Considering only concrete mathematical theories, one succeeds in constructing variants of calculi which are well adapted for deduction in these theories.

The goal of the present paper is to construct, taking into account general methods of eliminating cut rules (see, e.g., [1]) with the study of a specific axiom system of Raphael M. Robinson's arithmetic system [2, §41], a Gentzen type calculus not containing cut rules. The method of proving the equivalence of the constructed calculus with Robinson's arithmetic system enables us to give a new proof of the consistency of Robinson's system.

The construction in this paper of a calculus containing only one cut rule, equivalent to Robinson's system, might be of some independent interest.

AMS (MOS) subject classifications (1970). Primary 02C15, 02G99.

Since the eliminability of the interchange rule is easily proved (see, e.g., [3]) for Gentzen calculi, to facilitate the discussion, in all constructions of calculi in this paper, the postulates are such that they are applicable independently of the order of the formulas in the antecedents and succedents. In other words, every application of a postulate must remain an application of the same postulate if an arbitrary sequent is replaced by a sequent differing from it only by the order of the formulas.

The basic results of the paper with proofs sketched were published in [4].

The numbering of the theorems, lemmas, remarks, and displayed expressions runs through the entire article.

The author expresses his sincere gratitude to S. Ju. Maslov for his constant interest and valuable remarks during the writing of this paper and the preparation of it for the press. The author is also grateful to V. P. Orevkov for carefully reading the manuscript and for his remarks.

§1. Description of basic calculi

As an example of an initial calculus of R. M. Robinson's arithmetic system we shall consider the calculus (which we denote by $\mathbf{R}_{\text{init}}$) obtained in the manner described below from the calculus $\mathbf{R}_2$ [5], containing the following seven specific axioms:

$$1)\ \forall x\,(x = Pd\,(x')),\qquad 2)\ \forall x\,(x' \neq 0),$$
$$3)\ \forall x\,(x + 0 = x),\qquad 4)\ \forall xy\,(x + y' = (x + y)'),$$
$$5)\ \forall x\,(x \cdot 0 = 0),\qquad 6)\ \forall xy\,(x \cdot y' = (x \cdot y) + x),$$
$$7)\ \forall x\,(x = 0 \vee x = (Pd\,(x))'),$$

where Pd is a one-place functional symbol of the predecessor operation.

The axiom scheme of $\mathbf{R}_{\text{init}}$ is the scheme $\Gamma \to \Delta,\ r = r$.

The deduction rules of $\mathbf{R}_{\text{init}}$ split into two groups. The first group consists of the logical rules and rules for equality of Kanger type [6] which coincide with the corresponding rules of the calculus $\mathbf{R}_3$ (see [5]). The second group consists of the rules corresponding to specific axioms of $\mathbf{R}_2$, namely:

1) Instead of the specific axioms of the form $\forall x_1 \cdots x_m\ (t = \omega)$, i.e. axioms 1), 3), 4), 5) and 6), we introduce rules of the form

$$\frac{\begin{array}{c}\Gamma \to \left[\Delta \left|{}^{\alpha}_{\tilde{t}}\right.\right] \\ \alpha,\ \Gamma,\ \Delta \text{ satisfy Condition } (*)\end{array}}{\Gamma \to \left[\Delta \left|{}^{\alpha}_{\tilde{\omega}}\right.\right]}, \quad (1) \qquad \frac{\begin{array}{c}\Gamma \to \left[\Delta \left|{}^{\alpha}_{\tilde{\omega}}\right.\right] \\ \alpha,\ \Gamma,\ \Delta \text{ satisfy Condition } (*)\end{array}}{\Gamma \to \left[\Delta \left|{}^{\alpha}_{\tilde{t}}\right.\right]}, \quad (2)$$

where α is a free variable, Γ and Δ are lists of formulas, $T_1, \cdots, T_m$ are arbitrary terms, $\tilde{t}$ denotes $[t|^{x_1,\ldots,x_m}_{T_1,\ldots,T_m}]$ and $\tilde{\omega}$ denotes $[\omega|^{x_1,\ldots,x_m}_{T_1,\ldots,T_m}]$.

CONDITION (*): a) $\Gamma \to \Delta$ is a primary sequence (see [5]). b) α appears only in one member of the list of formulas Δ and this member is an atomic formula.

2) Instead of the specific axioms 2) and 7) we introduce the section rules

$$\frac{\Gamma \to \Delta,\ t' = 0}{\Gamma \to \Delta}, \ (3) \qquad \frac{t = 0,\ \Gamma \to \Delta; \quad t = (Pd\,(t))',\ \Gamma \to \Delta}{\Gamma \to \Delta}, \qquad (4)$$

respectively.

We shall call the formulas introduced explicitly in the cut rules (3) and (4) *side formulas* of these rules, and the term t a *proper term* of these rules. The side formulas of the remaining rules and also the principal formulas are defined in the usual way.

We note that the calculus $\mathbf{R}_{\text{init}}$ differs from the calculus $\mathbf{R}_3$ in [5] only by the absence of restrictions in rules (3) and (4) of $\mathbf{R}_{\text{init}}$ (these differences are of a technical character). The equivalence of $\mathbf{R}_2$ and $\mathbf{R}_{\text{init}}$ can be proven in exactly the same way as the equivalence of $\mathbf{R}_2$ and $\mathbf{R}_5$ (see [5]).

We now introduce the calculus $\mathbf{R}$ obtained from $\mathbf{R}_{\text{init}}$ by means of the following three transformations.

1. Replacement of rule (3) by the following axiom schemes:

$$t^{\overbrace{\prime \cdots \prime}^{n+1 \text{ times}}} = 0^{\overbrace{\prime \cdots \prime}^{n \text{ times}}},\ \Gamma \to \Delta, \qquad (5)$$

$$0^{\overbrace{\prime \cdots \prime}^{n \text{ times}}} = t^{\overbrace{\prime \cdots \prime}^{n+1 \text{ times}}},\ \Gamma \to \Delta. \qquad (6)$$

2. Replacement of rules of type (1) and (2), corresponding to the specific axioms 1), 3), 4), 5) and 6), by the rules (1′) and (2′), having the form of (1) and (2), but resolving substitution in the antecedent also, where in rules (1′) and (2′) the antecedent main formula is doubled. For example, one of the antecedent rules, corresponding to the specific axiom with matrix $t = \omega$, will have the form

$$\frac{\begin{gathered}\left[A \left|{}^{\alpha}_{t}\right.\right],\ \left[A \left|{}^{\alpha}_{\omega}\right.\right],\ \Gamma_1 \to \Delta \\ \alpha,\ \Gamma,\ \Delta \text{ satisfy Condition } (*)\end{gathered}}{\left[A \left|{}^{\alpha}_{t}\right.\right],\ \Gamma_1 \to \Delta},$$

and succedent rule

$$\frac{\begin{gathered}\Gamma \to \Delta_1,\ \left[A \left|{}^{\alpha}_{\omega}\right.\right] \\ \alpha,\ \Gamma,\ \Delta \text{ satisfy Condition } (*)\end{gathered}}{\Gamma \to \Delta_1,\ \left[A \left|{}^{\alpha}_{t}\right.\right]},$$

where the notation is the same as in rules (1) and (2) and in the antecedent variant of the rule $\Gamma \mathrel{\overline{\underline{\circ}}} [A|^{\alpha}_{t}], \Gamma_1$, and in the succedent variant of the rule $\Delta \mathrel{\overline{\underline{\circ}}} \Delta_1, [A|^{\alpha}_{\tilde{t}}]$.

3. Replacement of the rule for equality by rules in which the antecedent principal formula is doubled.

We obtain a new calculus $\mathbf{R}_{\text{fin}}$ (final) from the calculus $\mathbf{R}$ in the following way:

a) Replacement of the rules for equality and rules (1′) and (2′) by rules in which the antecedent and succedent principal rules are doubled; we shall denote the rules obtained from (1′) and (2′) by (1*) and (2*), respectively.

b) Replacement of rule (4) by the deduction rule

$$\frac{\Gamma_1 \to \Delta_1; \quad \Gamma_2 \to \Delta_2}{\Gamma \to \Delta}, \tag{7}$$

where for some term t the figure

$$\frac{\Gamma_1 \to \Delta_1}{\Gamma \to \Delta}$$

coincides with a rule (1*) or (2*) which corresponds to the equality $t = 0$. The right implication of rule (7) is obtained from $\Gamma \to \Delta$ analogously (using the equality $t = (Pd(t))'$). More precisely, suppose the formulas $[A|^{\alpha}_{\rho_1}]$ and $[B|^{\beta}_{\rho_2}]$ occur in $\Gamma\Delta$; then the formulas $[A|^{\alpha}_{\rho_1}]$, $[A|^{\alpha}_{\rho'_1}]$ and $[B|^{\beta}_{\rho_2}]$, occur in $\Gamma_1\Delta_1$, and $[A|^{\alpha}_{\rho_1}]$, $[B|^{\beta}_{\rho_2}]$ and $[B|^{\beta}_{\rho'_2}]$ occur in $\Gamma_2\Delta_2$, where $\rho_1 \mathrel{\overline{\underline{\circ}}} 0, t$; $\rho'_1 \mathrel{\overline{\underline{\circ}}} 0, t$; $\rho_2 \mathrel{\overline{\underline{\circ}}} t, (Pd(t))'$ and $\rho'_2 \mathrel{\overline{\underline{\circ}}} t, (Pd(t))'$, with $\rho_1 \not\mathrel{\overline{\underline{\circ}}} \rho'_1$ and $\rho_2 \not\mathrel{\overline{\underline{\circ}}} \rho'_2$; formulas A and B may coincide, whereas the remaining formulas of the lists $\Gamma\Delta$, $\Gamma_1\Delta_1$, and $\Gamma_2\Delta_2$ are the same.

Together with the schemes of rule (7) we can explicitly introduce rules obtained by the same scheme (16 rules altogether), but since listing them is cumbersome we shall limit ourselves to writing down some of them as an example:

$$\frac{[A|^{\alpha}_{0}], [A|^{\alpha}_{t}], [B|^{\beta}_{t}], \Gamma \to \Delta; \quad [A|^{\alpha}_{0}], [B|^{\beta}_{t}], [B|^{\beta}_{(Pd(t))'}], \Gamma \to \Delta}{[A|^{\alpha}_{0}], [B|^{\beta}_{t}], \Gamma \to \Delta},$$

$$\frac{[A|^{\alpha}_{0}], [A|^{\alpha}_{t}], \Gamma \to \Delta, [B|^{\beta}_{(Pd(t))'}]; \quad [A|^{\alpha}_{0}], \Gamma \to \Delta, [B|^{\beta}_{(Pd(t))'}], [B|^{\beta}_{t}]}{[A|^{\alpha}_{0}], \Gamma \to \Delta, [B|^{\beta}_{(Pd(t))'}]},$$

$$\frac{[B|^{\beta}_{t}], \Gamma \to \Delta, [A|^{\alpha}_{t}], [A|^{\alpha}_{0}]; \quad [B|^{\beta}_{t}], [B|^{\beta}_{(Pd(t))'}], \Gamma \to \Delta, [A|^{\alpha}_{t}]}{[B|^{\beta}_{t}], \Gamma \to \Delta, [A|^{\alpha}_{t}]},$$

$$\frac{\Gamma \to \Delta, [B|^{\beta}_{(Pd(t))'}], [A|^{\alpha}_{t}], [A|^{\alpha}_{0}]; \quad \Gamma \to \Delta, [B|^{\beta}_{(Pd(t))'}], [B|^{\beta}_{t}], [A|^{\alpha}_{t}]}{\Gamma \to \Delta, [B|^{\beta}_{(Pd(t))'}], [A|^{\alpha}_{t}]}.$$

In the sequel we shall call rules of the form (1), (2), (1′), (2′), (1*) and (2*) *axiomatic rules* for equality. We shall sometimes denote the axiomatic rules for equality (i') or (i*), corresponding to the specific axiom j, by $i \cdot j$, where $i = 1, 2$ and $j = 1, 3, 4, 5, 6$.

§ 2. Auxiliary calculus

In the sequel the expression $t^{[n]}$, where t is an arbitrary term and $n \geqslant 0$, will denote the term $t^{\overbrace{\prime \cdots \prime}^{n \text{ times}}}$. We introduce the following auxiliary calculus (denoted by **S**) whose deducible objects are objects of the form $t \Rightarrow 0^{[k]}$, where $\Rightarrow$ is a new formal symbol.

Axiom: $0 \Rightarrow 0$.

Deduction rules:

1) $\dfrac{p \Rightarrow 0^{[k]}}{p' \Rightarrow 0^{[k+1]}}$, where $k \geqslant 0$;

2) $\dfrac{p \Rightarrow 0^{[r]};\ q \Rightarrow 0^{[s]}}{p + q \Rightarrow 0^{[k]}}$, where $k = r + s,\ r \geqslant 0,\ s \geqslant 0$;

3) $\dfrac{p \Rightarrow 0^{[k+1]}}{P\dot{d}p \Rightarrow 0^{[k]}}$, where $k \geqslant 0$;

4) $\dfrac{p \Rightarrow 0}{v \cdot p \Rightarrow 0}$, where v is an arbitrary term;

5) $\dfrac{p \Rightarrow 0^{[r]};\ q \Rightarrow 0^{[s]}}{p \cdot q \Rightarrow 0^{[k]}}$,

where $k = r \cdot s$, $r \geqslant 0$ and $s \geqslant 1$.

For an arbitrary application of rule 4) all subterms of the term v will be called *special* terms of this application.

The expression $\vdash t \Rightarrow 0^{[k]}$ will denote that the object $t \Rightarrow 0^{[k]}$ is deducible in the calculus **S**. (It is easy to see that the deduction problem in **S** is solvable.)

REMARK 1. The object $t' \Rightarrow 0$, where t is an arbitrary term, is not deducible in **S**.

For arbitrary terms A, z and any number $k \geqslant 0$, we shall call the occurrence of the term z in $[A|_z^\alpha] \Rightarrow 0^{[k]}$ *special* if it arose in the deduction of the object $[A|_z^\alpha] \Rightarrow 0^{[k]}$ from the occurrence of some special term of some application of rule (4).

LEMMA 1. *For arbitrary terms A, z and any number $k \geqslant 0$, given the deduction of the object $[A|_z^\alpha] \Rightarrow 0^{[k]}$, in which the selected occurrence of the term z is not special, there can be found an $l \geqslant 0$ such that $\vdash z \Rightarrow 0^{[l]}$.*

PROOF. We denote by g the depth of occurrence of the term z in $[A|_z^\alpha]$ (see [2], § 26). We proceed by induction on g. If $g = 0$, then $[A|_z^\alpha \mathrel{\underline{\overline{\circ}}} z$ and $\vdash z \Rightarrow 0^{[k]}$. Let $g > 0$. We consider the lowest application of the rule in the deduction of the object $[A|_z^\alpha] \Rightarrow 0^{[k]}$ in **S**, and denote it by L. Then, applying the induction assumption in the premise and then applying rule L, we obtain that there exists an l such that $\vdash z \Rightarrow 0^{[l]}$.

LEMMA 2. *For an arbitrary term z and natural numbers l and k, both of the objects $z \Rightarrow 0^{[k]}$ and $z \Rightarrow 0^{[l]}$ are deducible in **S** if and only if $l = k$.*

The lemma is proved by induction on the length of the term z.

LEMMA 3. *For arbitrary terms* A, x *and* z *and any natural number* k, *if* $\vdash [A|_z^\alpha] \Rightarrow 0^{[k]}$ *and there exists an* $l \geqslant 0$ *such that* $\vdash z \Rightarrow 0^{[l]}$ *and* $\vdash x \Rightarrow 0^{[l]}$, *then* $\vdash [A|_x^\alpha] \Rightarrow 0^{[k]}$.

The proof is analogous to that of Lemma 1; it is carried out by induction on the depth of occurrence of the term z in $[A|_z^\alpha]$ using Lemma 2 in the initial step of the induction.

The *axiomatic deduction of the term* t_0 from the term t_n is a figure D consisting of only n $(n \geqslant 0)$ applications of axiomatic rules for equality and such that for some A and α the principal formula of an arbitrary application of the rule in the figure D has the form $[A|_{t_{i-1}}^\alpha]$ $(i = 1, \cdots, n)$ and is an ancestor of the occurrences of a formula of the form $[A|_{t_0}^\alpha]$ in a lower sequent of the figure D.

It is easy to see that if it is possible to construct an axiomatic deduction of the term t_0 from the term t_n, then it is also possible to construct an axiomatic deduction of t_n from t_0. In the sequel, in writing down the axiomatic deduction of t_0 from t_n, instead of an arbitrary step i of the deduction D of the form

$$\frac{[A|_{t_i}^\alpha]}{[A|_{t_{i-1}}^\alpha]}$$

we shall only write the relevant terms, i.e.

$$\frac{t_i}{t_{i-1}},$$

and we shall represent the entire deduction as

$$D\left\{\begin{array}{c} \underline{t_n} \\ \underline{\cdots} \\ t_0 \end{array}\right. .$$

THEOREM 1. *For an arbitrary term* t *and natural number* k, *a necessary and sufficient condition that* $\vdash t \Rightarrow 0^{[k]}$ *in the calculus* **S** *is that it be possible to construct an axiomatic deduction of* t *from the term* $0^{[k]}$.

PROOF. *Necessity.* Let $\vdash t \Rightarrow 0^{[k]}$. We shall show that it is then possible to construct the term t from the term $0^{[k]}$ by axiomatic deduction. We use induction on the height of the deduction V of the object $t \Rightarrow 0^{[k]}$. The initial step of the induction is trivial since then $t \underset{\circ}{=} 0^{[k]}$, $k = 0$.

Induction step. We denote by i the lowest application of the deduction rule in the deduction V in the calculus **S**. We consider cases depending on the form of the rule i.

Case 1. i is an application of rule 2), i.e. the deduction V has the form

$$V\left\{\frac{\dfrac{\cdots}{p\Rightarrow 0^{[r]}};\quad \dfrac{\cdots}{q\Rightarrow 0^{[s]}}}{p+q\Rightarrow 0^{[r+s]}}\ 2).\right.$$

According to the induction assumption, it is possible to construct axiomatic deductions D_1 and D_2 of the terms p and q from the terms $0^{[r]}$ and $0^{[s]}$, respectively. We construct the axiomatic deduction sought in the following way, using the deductions D_1, D_2 and $0^{[r+s]} \underset{\circ}{=} 0^{[r][s]}$:

$$\begin{array}{lcl}
 & 0^{[r][s]} & \\ \cline{2-2}
D_1 & \cdots & \\ \cline{2-2}
 & p^{[s]} & \\ \cline{2-2}
 & (p+0)^{[s]} & 2.3 \\ \cline{2-2}
 & (p+0')^{[s-1]} & 2.4 \\ \cline{2-2}
 & \cdots & \\ \cline{2-2}
 & (p+0^{[s-1]})' & 2.4 \\ \cline{2-2}
 & p+0^{[s]} & 2.4. \\ \cline{2-2}
D_2 & \cdots & \\ \cline{2-2}
 & p+q &
\end{array}$$

Case 2. i is an application of rule 5), i.e. the deduction V has the form

$$V\left\{\frac{\dfrac{\cdots}{p\Rightarrow 0^{[r]}};\quad \dfrac{\cdots}{q\Rightarrow 0^{[s]}}}{p\cdot q\Rightarrow 0^{[r\cdot s]}}\ 5).\right.$$

According to the induction assumption, we can construct the axiomatic deductions D_1 and D_2 of the terms p and q from $0^{[r]}$ and $0^{[s]}$, respectively. Before constructing the desired axiomatic deduction, we define the following figure ϑ_i $(i = 1, \cdots, s)$:

$$\begin{array}{cl}
(0^{[r]}\,.\,0^{[i-1]})^{[r\,\cdot\,(s-i+1)]} & \\ \hline
(((0^{[r]}\,.\,0^{[i-1]})+0)^{[r]})^{[r\,\cdot\,(s-i)]} & 2.3 \\ \hline
\cdots & \text{rule 2.4 } (r-1) \text{ times} \\ \hline
(((0^{[r]}\,.\,0^{[i-1]})+0^{[r-1]})')^{[r\,\cdot\,(s-i)]} & \\ \hline
((0^{[r]}\,.\,0^{[i-1]})+0^{[r]})^{[r\,\cdot\,(s-i)]} & 2.4 \\ \hline
(0^{[r]}\,.\,0^{[i]})^{[r\,\cdot\,(s-i)]} & 2.6.
\end{array}$$

We construct the desired axiomatic deduction in the following way:

$$\begin{array}{lcl}
 & 0^{[r\,\cdot\,s]} & \\ \cline{2-2}
 & (0^{[r]}\,.\,0)^{[r\,\cdot\,s]} & 2.5 \\ \cline{2-2}
\vartheta_1 & \cdots & \\ \cline{2-2}
 & (0^{[r]}\,.\,0^{[s-(s-1)]})^{[r\,\cdot\,(s-1)]} &
\end{array}$$

$$\begin{array}{l} \vartheta_s \left\{ \begin{array}{c} \ldots \\ \hline (0^{[r]} \cdot 0^{[s-1]})^{[r]} \\ \ldots \\ \hline \end{array} \right. \\ D_2 \left\{ \begin{array}{c} 0^{[r]} \cdot 0^{[s]} \\ \ldots \\ \hline \end{array} \right. \\ D_1 \left\{ \begin{array}{c} 0^{[r]} \cdot q \\ \ldots \\ \hline p \cdot q \end{array} \right. \end{array} .$$

The cases when i is an application of rule 1), 3) or 4) are considered analogously. This proves the necessity.

Sufficiency. Suppose it is possible to construct an axiomatic deduction D of the term t from the term $0^{[k]}$. We then show that $\vdash t \Rightarrow 0^{[k]}$. We use induction on the height h of the axiomatic deduction D.

Initial step: $h = 1$. In this case $t \overline{\overline{\circ}}\, 0^{[k]}$ and the deduction of the object $0^{[k]} \Rightarrow 0^{[k]}$ is the following:

$$\left.\begin{array}{c} 0 \Rightarrow 0 \\ \ldots \\ \hline 0^{[k-1]} \Rightarrow 0^{[k-1]} \end{array}\right\} \text{rule 1) applied } (k-1) \text{ times}$$
$$\frac{0^{[k-1]} \Rightarrow 0^{[k-1]}}{0^{[k]} \Rightarrow 0^{[k]}}\ 1).$$

Induction step: $h > 1$. We consider the last step of the deduction D. We denote by i the lowest application of the axiomatic rule for equality in D, i.e. the end of D has the form

$$\frac{[t|^{\gamma}_{v}]}{[t|^{\gamma}_{\omega}]}\, i.$$

According to the induction assumption, we can construct the deduction V of the object $[t|^{\gamma}_{v}] \Rightarrow 0^{[k]}$ in the calculus **S**. We consider two cases.

Case 1. Suppose the selected occurrence of the term v in $[t|^{\gamma}_{v}] \Rightarrow 0^{[k]}$ is special. It follows from the form of the postulates of the calculus **S** that by replacing all occurrences of v in the given deduction V by the term ω we obtain the deduction of the object $[t|^{\gamma}_{\omega}] \Rightarrow 0^{[k]}$.

Case 2. Suppose that in the deduction V all applications of rule 4) are such that the selected occurrence of v is not special. According to Lemma 1 there exists an $l \geqslant 0$ such that we can construct a deduction V^* of the object $v \Rightarrow 0^{[l]}$ in the calculus **S**. We shall show that it is possible to construct a deduction V' of the object $\omega \Rightarrow 0^{[l]}$. By Lemma 3 it follows from this that $\vdash [t|^{\gamma}_{\omega}] \Rightarrow 0^{[k]}$. We must consider various cases depending on the form of rule i in D. We consider some of them.

Subcase 2.1. i is an application of rule 1.5, i.e. $v \overline{\overline{\circ}}\, p \cdot 0$ and $\omega \overline{\overline{\circ}}\, 0$. The end of the deduction V^* has the form

$$\frac{0 \Rightarrow 0^{[l]}}{p \cdot 0 \Rightarrow 0^{[l]}}\ 4).$$

In this case $l = 0$, since otherwise the premise of rule 4) will not be deducible. The desired deduction V' is the deduction of the object $0 \Rightarrow 0$.

We note that the end of the deduction V^* cannot have the form

$$\frac{p \Rightarrow 0^{[r]};\quad 0 \Rightarrow 0^{[s]}}{p \cdot 0 \Rightarrow 0^{[l]}}\,5),$$

where $l = r \cdot s$, $r \geqslant 0$, $s \geqslant 1$, because in this case the right premise is not deducible.

Subcase 2.2. i is an application of rule 2.6, i.e. $v \mathrel{\overline{\overline{\circ}}} (p \cdot q) + p$ and $\omega_0 \mathrel{\overline{\overline{\circ}}} p \cdot q'$. In this case, the deduction V^* can end in the following two ways.

First way:

$$V^*\left\{\frac{\dfrac{V_1\left\{\dfrac{\cdots}{p \Rightarrow 0^{[r_1]}}\right.;\quad V_2\left\{\dfrac{\cdots}{q \Rightarrow 0^{[r_2]}}\right.}{p \cdot q \Rightarrow 0^{[r]};}\,5)\qquad \dfrac{\cdots}{p \Rightarrow 0^{[s]}}}{(p \cdot q) + p \Rightarrow 0^{[l]}}\,2),\right.$$

where $l = r + s$; $r = r_1 \cdot r_2$, $r \geqslant 0$; $r_1 \geqslant 0$; $r_2 > 0$, and, according to Lemma 2, $r_1 = s$. We construct the desired deduction V' in the following way:

$$\frac{V_1\left\{\dfrac{\cdots}{p \Rightarrow 0^{[s]}}\right.;\qquad \dfrac{V_2\left\{\dfrac{\cdots}{q \Rightarrow 0^{[r_2]}}\right.}{q' \Rightarrow 0^{[r_2+1]}}\,1)}{p \cdot q' \Rightarrow 0^{[l]}}\,5),$$

where $l = s \cdot (r_2 + 1) = s \cdot r_2 + s = r_1 \cdot r_2 + s = r + s$.

Second way:

$$V^*\left\{\frac{\dfrac{V_1\left\{\dfrac{\cdots}{q \Rightarrow 0}\right.}{p \cdot q \Rightarrow 0^{[r]};}\,4)\qquad V_2\left\{\dfrac{\cdots}{p \Rightarrow 0^{[s]}}\right.}{(p \cdot q) + p \Rightarrow 0^{[l]}}\,2),\right.$$

where $l = r + s$, $r = 0$ and $s \geqslant 0$, i.e., $l = s$.

We construct the desired deduction V' as follows:

$$\frac{V_2\left\{\dfrac{\cdots}{p \Rightarrow 0^{[s]}}\right.;\qquad \dfrac{V_1\left\{\dfrac{\cdots}{q \Rightarrow 0}\right.}{q' \Rightarrow 0^{[1]}}\,1)}{p \cdot q' \Rightarrow 0^{[l]}}\,5),$$

where $l = s \cdot 1 = s$.

The remaining cases are considered analogously.

This proves the sufficiency.

§3. Equivalence of the calculi $\mathbf{R}_{\mathrm{init}}$ and $\mathbf{R}$

Following M. G. Rogava [5], we shall call an application of a deduction rule *antecedent* (*succedent*) if the succedent (antecedent) of the conclusion of the rule coincides with the succedent (antecedent) of each premise. We shall call a deduction *antecedent* (*succedent*) if all applications of the rules in it are antecedent (succedent). Application of the rule for equality or of the axiomatic rule for equality will be said to be left-sided (right-sided) if the principal formula of this application is equality and in this application of this rule only the left (right) side of this equality undergoes change.

In the sequel, we shall denote by $\mathbf{J}$ the auxiliary calculus obtained from the calculus $\mathbf{R}$ by removal of logical rules and rule (4). The calculus $\mathbf{J}'$ is constructed from the calculus $\mathbf{R}_{\mathrm{init}}$ by means of transformations given in the definition of $\mathbf{J}$ and the elimination of rule (3).

We call a deduction in $\mathbf{J}$ or $\mathbf{J}'$ *left-sided* (*right-sided*) if no application of the rules in this deduction is right-sided (left-sided). We shall call a deduction in $\mathbf{J}$ or $\mathbf{J}'$ *one-sided* if it is either right-sided or left-sided.

We shall call an axiom of the form $\Gamma \to \Delta,\ r = r$ a *succedent axiom*, and an axiom of the form (5) or (6) an *antecedent axiom*; moreover, we shall call the formula which is explicitly selected in the scheme of the axiom the *principal formula of the axiom.*

LEMMA 4. *Given a succedent left-sided deduction V in the calculus $\mathbf{R}_{\mathrm{init}}$, whose last step consists in the application of rule* (3), *with deduction in the calculus $\mathbf{J}'$ placed on the premise of this application, one can construct a deduction V' in $\mathbf{J}$ with the same last sequent.*

PROOF. Deduction V has the form

$$V\left\{\begin{array}{l} \left.\begin{array}{c} a=b,\ \Gamma\to 0=0 \\ \cdots \\ \hline a=b,\ \Gamma\to [A|^{\alpha}_{b}]=0 \end{array}\right\} D \\ \hline a=b,\ \Gamma\to [A|^{\alpha}_{a}]=0 \quad\quad \to = \\ \quad\quad \cdots \\ \hline a=b,\ \Gamma\to t'=0 \\ \hline a=b,\ \Gamma\to \quad\quad (3), \end{array}\right.$$

i.e. it must contain at least one application of the rule for equality. In fact, it follows from Remark 1 that the object $t' \Rightarrow 0$ is not deducible in $\mathbf{S}$, and by Theorem 1 this means that it is impossible to construct an axiomatic deduction of the term t' from the term 0. Suppose that the explicitly described application of the rule for equality in the deduction V is the highest application of the rule for equality. Then there exists an axiomatic deduction D of the term $[A|^{\alpha}_{b}]$ from the term 0 and, by

Theorem 1, $\vdash [A|_b^\alpha] \Rightarrow 0$ in **S**. Without loss of generality we shall assume that the selected occurrence of the term b in the term $[A|_b^\alpha]$ is not special (otherwise, $\vdash [A|_a^\alpha] \Rightarrow 0$, i.e. the given deduction V can be reconstructed into a deduction not containing the selected application of the rule for equality; the number of applications of the rule for equality is decreased, but, according to Theorem 1 and Remark 1, the reconstructed deduction contains at least one application of the rule for equality). Since the conditions of Lemma 1 are satisfied, there exists, by Lemma 1, an $l \geqslant 0$ such that $\vdash b \Rightarrow 0^{[l]}$. In accordance with Theorem 1, we construct with respect to the deduction $b \Rightarrow 0^{[l]}$ in the calculus **S** the axiomatic deduction D' of the term b from the term $0^{[l]}$. We construct the desired deduction V' as follows:

$$V'\left\{\begin{array}{l} D_1\left\{\begin{array}{l} \dfrac{0^{[l]}=t^{[l+1]},\ \Gamma,\ \Sigma,\ a=0^{[l]},\ \Sigma_1,\ a=t^{[l+1]}\to}{a=t^{[l+1]},\ \Gamma,\ \Sigma,\ a=0^{[l]},\ \Sigma_1\to}\ {=}\!\to \\ \qquad\qquad \dots \\ \hline \end{array}\right. \\ D'\left\{\begin{array}{l} a=0^{[l]},\ \Gamma,\ \Sigma\to \\ \qquad \dots \\ \hline a=b,\ \Gamma\to \end{array}\right., \end{array}\right.$$

where D_1 consists of applications of the rules which carry the term t' into the term 0, constructed with respect to the deduction V; Σ and Σ_1 are lists of formulas arising from doubling the principal formulas upon application of the rules for equality and rules (1′) and (2′); the auxiliary formula in the explicitly selected application of the rule for equality is the formula $a = 0^{[l]}$. The first sequent of the deduction V' is an axiom of the form (6). This completes the proof of the lemma.

LEMMA 6. *Given the deduction V in the calculus $\mathbf{R}_{\text{init}}$ possessing the properties listed in the formulation of Lemma* 4, *one can construct a deduction V_1 in the calculus* **J** *with the same last sequent and moreover such that the principal formula of an antecedent axiom can be the ancestor of a formula of the form $t = (Pd(t))'$, if and only if there exists a succedent $k \geqslant 0$ such that $\vdash t \Rightarrow 0^{[k]}$ in the calculus* **S**.

PROOF. To prove the lemma we use figures of the deduction V in $\mathbf{R}_{\text{init}}$ and of the deduction V' in **J** which were written out in the proof of Lemma 4. As is clear from the reconstruction made in the deduction V, the principal formula of the antecedent axiom in the deduction V' is a predecessor of the formula $a = b$ which occurs in the last sequent of V and is an auxiliary formula of the application considered of the rule for equality in V. We consider two cases.

Case 1. Suppose the auxiliary formula $a = b$ of the application considered of the rule for equality in V is not a formula of the type $t = (Pd(t))'$. Then the deduction V', constructed in the proof of Lemma 4, is the one sought.

Case 2. Suppose the auxiliary formula $a = b$ of the application considered of the rule for equality in V is a formula of the form $t = (Pd(t))'$. Since one can attain the situation that the selected occurrence of the term b in the term $[A|_b^\alpha]$ is not special (see the proof of Lemma 4), it follows from Lemma 1 that there exists a $k \geqslant 0$ such that $\vdash b \Rightarrow 0^{[k]}$ in $\mathbf{S}$. In our case $b \mathrel{\underline{\overline{\circ}}} t$ or $b \mathrel{\underline{\overline{\circ}}} (Pd(t))'$.

Let $b \mathrel{\underline{\overline{\circ}}} t$. Then there exists a $k \geqslant 0$ such that $\vdash t \Rightarrow 0^{[k]}$ in $\mathbf{S}$.

Let $b \mathrel{\underline{\overline{\circ}}} (Pd(t))'$. Then there exists a $k > 0$ such that $\vdash (Pd(t))' \Rightarrow 0^{[k]}$ in $\mathbf{S}$ (since the object $(Pd(t))' \Rightarrow 0$ is not deducible in $\mathbf{S}$; see Remark 1). It follows from the reversibility of postulate 1) of $\mathbf{S}$ that the object $(Pd(t))' \Rightarrow 0^{[k-1]}$ is deducible, and it follows from the reversibility of postulate 3) that $\vdash t \Rightarrow 0^{[k]}$.

Deduction V_1 is constructed from the given deduction V in the same way as in the proof of Lemma 4 (in this case we obtain that the principal formula of the antecedent axiom is the predecessor of a formula of the form $t = (Pd(t))'$ and there exists a $k \geqslant 0$ such that $\vdash t \Rightarrow 0^{[k]}$). This completes the proof of the lemma.

We shall say that a deduction in $\mathbf{R}_{\text{init}}$ (in $\mathbf{R}$) is *regular* if in it all applications of rules (3) and (4) (rule (4), respectively) occur above applications of logical rules, but below all applications of rules for equality and axiomatic rules for equality.

Using the well developed apparatus of the reconstruction of a deduction (see, e.g., [7]), one can show that an arbitrary deduction in $\mathbf{R}_{\text{init}}$ (in $\mathbf{R}$) can be reconstructed into a regular deduction in the same calculus (see [1], Lemma 1).

We shall call a deduction in $\mathbf{R}_{\text{init}}$ (in $\mathbf{R}$) *weeded* if one of the side formulas (both side formulas in the case of a bifurcating rule) of an arbitrary application of the rule in a given deduction satisfies one of the following conditions:

1) It is a successor of the principal formula of some axiom.

2) It is a successor of an auxiliary formula of some application of the rule for equality.

It is clear from the definition that a weeded deduction does not contain "superfluous" applications of rules, but it may contain "superfluous" formulas.

LEMMA 6. *An arbitrary deduction in* $\mathbf{R}_{\text{init}}$ *(in* $\mathbf{R}$*) can be reconstructed into a weeded deduction in the same calculus with the same last sequent.*

The lemma is proved analogously to the proofs of Lemmas 2.2 and 2.3 in [8].

LEMMA 7. *A weeded regular deduction in* $\mathbf{R}_{\text{init}}$ *can be reconstructed into a deduction of the same sequent in the same calculus in which only applications of the rules for equality and rules* (1) *and* (2) *occur above an arbitrary application of rule* (3).

Lemma 7 is a particular case of Lemma 2 in [1] for the concrete calculus $\mathbf{R}_{\text{init}}$, and it follows in an obvious way from the weededness of the deduction.

LEMMA 8. *A deduction in* $\mathbf{R}_{\text{init}}$ *whose last step consists in an application of rule* (3) *whose premise is a deduction in the calculus* $\mathbf{J}'$ *can be reorganized into a succedent one-sided deduction with the same last sequent in* $\mathbf{R}_{\text{init}}$.

This lemma is proved in exactly the same way as were Lemmas 2 and 3 in [5], first using induction on the number of antecedent applications of the rules in the given deduction, and then induction on the number of right-sided applications of the rules for constructing a left-sided deduction and conversely.

We denote by $\mathbf{R}'$ the calculus obtained from the calculus $\mathbf{J}$ by adjoining rule (3).

LEMMA 9. *Suppose given a deduction of the sequent* $\mathbf{S}$ *in the calculus* $\mathbf{J}$, *whose first sequent is an antecedent axiom. Then one can construct a deduction of* $\mathbf{S}$ *in the calculus* $\mathbf{R}'$ *whose first sequent is a succedent axiom.*

PROOF. Without loss of generality we may assume that the given deduction has the form

$$V\left\{\frac{\begin{array}{c} t^{[n+1]}=0^{[n]},\ \Gamma,\ \Gamma_1\to\Delta' \\ \cdots \end{array}}{\Gamma\to\Delta}\right.,$$

where Γ_1 are the formulas appearing in the premises of applications of the antecedent rules of $\mathbf{J}$. We construct the desired deduction as follows:

$$\begin{array}{ll}
\dfrac{t^{[n+1]}=0^{[n]},\ \Gamma,\ \Gamma_1\to\Delta',\ 0=0}{t^{[n+1]}=0^{[n]},\ \Gamma,\ \Gamma_1\to\Delta',\ Pd\,(0')=0} & 1.1 \\
\left.\dfrac{\cdots}{t^{[n+1]}=0^{[n]},\ \Gamma,\ \Gamma_1\to\Delta',\ \underbrace{Pd\,(\ldots(Pd}_{n-1\text{ times}}\,(0^{[n-1]}))\ldots)=0}\right\} & 1.1\ (n-2)\text{ times} \\
\dfrac{}{t^{[n+1]}=0^{[n]},\ \Gamma,\ \Gamma_1\to\Delta',\ \underbrace{Pd\,(\ldots(Pd}_{n\text{ times}}\,(0^{[n]}))\ldots)=0} & 1.1 \\
\dfrac{}{t^{[n+1]}=0^{[n]},\ \Gamma,\ \Gamma_1\to\Delta',\ \underbrace{Pd\,(\ldots(Pd}_{n\text{ times}}\,(t^{[n+1]}))\ldots)=0} & R_1 \\
\left.\dfrac{\cdots}{t^{[n+1]}=0^{[n]},\ \Gamma,\ \Gamma_1\to\Delta',\ Pd\,(t^{[2]})=0}\right\} & 2.1\ (n-1)\text{ times} \\
\dfrac{}{t^{[n+1]}=0^{[n]},\ \Gamma,\ \Gamma_1\to\Delta',\ t'=0} & 2.1 \\
\dfrac{}{t^{[n+1]}=0^{[n]},\ \Gamma,\ \Gamma_1\to\Delta'} & (3)
\end{array}$$

$$V\left\{\frac{\begin{array}{c} t^{[n+1]}=0^{[n]},\ \Gamma,\ \Gamma_1\to\Delta' \\ \cdots \end{array}}{\Gamma\to\Delta}\right.,$$

where R_1 is application of the succedent rule for equality with the auxiliary formula $t^{[n+1]}=0^{[n]}$.

LEMMA 10. *A weeded deduction in the calculus* $\mathbf{R}'$ *can be reconstructed into a succedent deduction in the same calculus and with the same last sequent.*

PROOF. Suppose the first sequent of the given deduction in $\mathbf{R}'$ is an antecedent axiom. Since the given deduction is weeded, it contains no applications of rule (3), i.e. this deduction is in **J**. According to Lemma 9 this deduction can be reconstructed into a deduction in $\mathbf{R}'$ whose first sequent is a succedent axiom.

To reconstruct a deduction whose first sequent is a succedent axiom into a succedent deduction we use induction on the number of antecedent applications of the rules (cf. Lemma 8).

THEOREM 2. *The calculi* $\mathbf{R}_{\text{init}}$ *and* $\mathbf{R}$ *are equivalent.*

PROOF. In accordance with Lemmas 6 and 7, we reconstruct an arbitrary deduction in $\mathbf{R}_{\text{init}}$ into a regular cleansed deduction in which there appear only applications of the rules for equality and rules (1) and (2) above an arbitrary application of rule (3).

In accordance with Lemmas 8 and 4, we reconstruct all parts of deductions whose last step is application of rule (3) into deductions in the calculus **J**. Replacing in the deduction thus obtained all applications of rules for equality and rules (1) and (2) by the corresponding rules of the calculus **R**, we obtain a deduction in **R**.

To obtain a deduction in $\mathbf{R}_{\text{init}}$ we reconstruct a deduction in **R** into a regular weeded (according to Lemma 6) deduction. In the constructed deduction we select the parts of the deduction which are deductions in **J**, and by the same token also in $\mathbf{R}'$, and are such that the last sequent of a given part of the deduction is not a premise of an application of the rule for equality or of rule (1′) or (2′). Applying Lemma 10, we reconstruct all such parts of deductions into succedent deductions in $\mathbf{R}'$. Besides applications of logical rules and rules (3) and (4), the constructed deduction contains only succedent applications of the rules for equality and axiomatic rules for equality which coincide with the analogous rules of $\mathbf{R}_{\text{init}}$. It follows from the construction of the deduction that it contains no antecedent axioms, i.e. the constructed deduction is a deduction in $\mathbf{R}_{\text{init}}$.

§4. Normal deductions in the calculus R

Below, we shall prove the equivalence of the calculi **R** and $\mathbf{R}_{\text{fin}}$. In proving that it is possible to reconstruct a deduction in **R** into a deduction in $\mathbf{R}_{\text{fin}}$ a central role is played by replacement of an application of rule (4) by an application of rule (7). To this end, it is more convenient to reconstruct an arbitrary deduction in **R** into a deduction of some special type in which the side formulas of an application of rule (4) are not subjected to change. We shall prove a number of theorems and auxiliary lemmas on specialization of deduction in **R**.

Suppose given two applications of rule (4), one of which appears directly above the other:

$$\mathfrak{D}_1\left\{\dfrac{\dfrac{r=0,\ t=0,\ \Gamma\to\Delta;\quad r=(Pd\,(r))',\ t=0,\ \Gamma\to\Delta}{t=0,\ \Gamma\to\Delta;}\,(4)\quad t=(Pd\,(t))',\ \Gamma\to\Delta}{\Gamma\to\Delta}\right. .$$

We shall say that the configuration $\mathfrak{D}_2$ is obtained from $\mathfrak{D}_1$ by the *operation of interchanging* applications of rule (4) if it has the form

$$\mathfrak{D}_2 \left\{ \dfrac{\dfrac{t=0,\ r=0,\ \Gamma\to\Delta;\ t=(Pd\,(t))',\ r=0,\ \Gamma\to\Delta}{r=0,\ \Gamma\to\Delta}\,(4);\quad \dfrac{t=0,\ r=(Pd\,(r))',\ \Gamma\to\Delta;\ t=(Pd\,(t))',\ r=(Pd\,(r))',\ \Gamma\to\Delta}{r=(Pd\,(r))',\ \Gamma\to\Delta}\,(4)}{\Gamma\to\Delta}\,(4). \right.$$

REMARK 2. To every upper sequent of $\mathfrak{D}_1$ there corresponds an upper sequent in $\mathfrak{D}_2$, which either coincides precisely with the sequent of $\mathfrak{D}_1$, or differs from it only by the presence of "superfluous" formulas.

Applying the interchange operation instead of an application of rule (4) which appears lower in $\mathfrak{D}_1$, we obtain two applications of rule (4) with the same special term in $\mathfrak{D}_2$. We shall call the upper applications of rule (4) in $\mathfrak{D}_2$, obtained as a result of application of the interchange operation, *successors* of the lower application of rule (4) of $\mathfrak{D}_1$. If in $\mathfrak{D}_1$ the upper application of rule (4) appears above the right premise of the lower application of rule (4), then the definitions of the interchange operation and successor are completely analogous.

Suppose that in $\mathfrak{D}_2$ there are other applications of rule (4) above the successor of the lower application of rule (4); then the interchange operation can also be applied to them. Moreover, the successors of successors of a lower (for configuration $\mathfrak{D}_1$) application of rule (4) will also be called *successors* of this application of rule (4). We note that the figure constructed as a result of a repeated application of the interchange operation relative to some given figure in which above the lower application of rule (4) there are other applications of rule (4) will contain some successors of a lower application of rule (4) (one more than the number of applications of the interchange operation).

The following lemmas on the interchange of applications of rule (4) in a given deduction hold.

LEMMA 11. *Suppose given a regular deduction V in the calculus* **R**. *For an arbitrary choice of application of rule* (4) *one can construct a sequence of applications of the interchange operation, yielding a deduction V' in which the following conditions are satisfied*:

1) *There are no other applications of rule* (4) *above any successor of a selected application of rule* (4).

2) *The predecessors of side formulas of an application of rule* (4) *in V' change in the same way as the predecessors of side formulas of the corresponding applications of rule* (4) *in V.*

PROOF. Since V is regular, only applications of rule (4) can appear between two applications of rule (4), and above the highest applications of rule (4) in each branch there appear only applications of rules for equality and rules (1) and (2). We shall prove the lemma by induction on the number n of applications of rule (4) appearing above a selected application of rule (4).

If $n = 0$, then the deduction V is the one sought.

Let $n > 0$. We consider the lowest left application of rule (4) appearing higher than the selected application of rule (4). Apply the interchange operation to these applications of rule (4). It follows from Remark 2 and the allowability of refinement that after application of the interchange operation we obtain a deduction tree with the same last sequent, and condition 2) in the formulation of the lemma will be satisfied in the constructed deduction. The induction hypothesis is applicable to each of the deductions whose last steps are the successor of the selected application of rule (4) inasmuch as the induction parameter has been decreased. The deduction V' has been constructed.

This completes the proof of the lemma.

We note that if in the deduction V considered in Lemma 11 there are applications of rule (4) above both premises of the selected application of rule (4), then in the deduction V' the number of applications of rule (4) which are not successors of the selected application of rule (4) will be greater than in the initial deduction. Indeed, if in figure $\mathfrak{D}_1$ we had one application of rule (4) with special term q above the right premise of the selected application of rule (4), then in $\mathfrak{D}_2$ we have an application of rule (4) with the same special term q above the right premise of each of the successors of the lower application of rule (4).

The following lemma is a modification of Lemma 11; we consider the possibility of reconstructing a deduction into a deduction in which there are no further applications of rule (4) above one of the premises of successors of the selected application of rule (4).

LEMMA 12. *Let V be a given regular deduction in the calculus* **R**. *For any selection of application of rule* (4) *it is possible to construct a sequence of applications of the interchange operation yielding a deduction V' in which the following conditions are satisfied*:

1) *There are no further applications of rule* (4) *above the left* (*right*) *premise of an arbitrary successor of the selected application of rule* (4).

2) *The predecessors of side formulas of applications of rule* (4) *change in the same way as the predecessors of side formulas of the corresponding applications of rule* (4) *in V.*

3) *The number of applications of rule* (4) *which are not successors of the selected application of rule* (4) *and appearing below applications of successors of the selected application of rule* (4) *in V' is not greater than the number of applications of rule* (4) *in V.*

PROOF. The lemma is proved by induction on the number n of applications of rule (4) appearing above the left (right) premise of the selected application of rule (4). For definiteness we consider the case of a left premise.

If $n = 0$, the given deduction V is the one sought.

Let $n > 0$. We consider the lowest application of rule (4) appearing above a left premise of the selected application of rule (4). The possibility of constructing from a given deduction V a deduction in which conditions 1) and 2) of the lemma are satisfied is proved the same way as in Lemma 11. Condition 3) of the lemma will be satisfied in the deduction thus constructed inasmuch as in the induction step the successors of the selected application of rule (4) (in particular, the selected application of rule (4) itself) rises only above the applications of rule (4) appearing above the left premise. It follows from the form of the configurations $\mathfrak{D}_1$ and $\mathfrak{D}_2$ that the total number of applications of rule (4) appearing above left premises of successors of the selected application of rule (4) is not greater than the number of applications of rule (4) appearing above a left premise of the selected application of rule (4). This completes the proof of the lemma.

LEMMA 13. *Given the weeded deduction V of the sequent $\Gamma \to \Delta$ in the calculus* **R**, *whose last step consists of the application of a rule* (4) *such that for a special term t of this application there exists an $l \geqslant 0$ such that $\vdash t \Rightarrow 0^{[l]}$ in the calculus* **S** *and in the case $l = 0$ ($l > 0$) above a left (respectively, right) premise of this application there appears a deduction in the calculus* **J**, *one can construct a deduction of the sequent $\Gamma \to \Delta$ in* **J**.

PROOF. The end of the given deduction V has the form

$$V\left\{\frac{V_1\left\{\frac{\cdots}{t=0,\ \Gamma\to\Delta;}\right. \qquad V_2\left\{\frac{\cdots}{t=(Pd\,(t))',\ \Gamma\to\Delta}\right.}{\Gamma\to\Delta}\right. (4).$$

It follows from the condition of the lemma that there exists an $l \geqslant 0$ such that the object $t \Rightarrow 0^{[l]}$ is deducible in **S**. According to Lemma 1 we can construct an axiomatic deduction of the term t from $0^{[l]}$ ($l \geqslant 0$). We consider two cases.

Case 1. $l = 0$, i.e. $\vdash t \Rightarrow 0$ in **S**. We then reconstruct the left branch of the deduction, which is a deduction in **J**, in the following way. If the first sequent of the deduction V_1 is a succedent axiom, then we reconstruct V_1 into the succedent deduction V_1' in **J** in accordance with Lemma 10. And if the first sequent of V_1 is an antecedental axiom, then we reconstruct V_1 into the succedent deduction V_1' in the calculus **R'** in accordance with Lemmas 9 and 10. Since the given deduction V is weeded, in the deduction V_1' of the left premise the side formula $t = 0$ must be used as an auxiliary formula for equality. We shall show that V_1' can be reconstructed into a deduction V_1^* which does not contain applications of rules for equality, whose auxiliary formula is $t = 0$. We replace an arbitrary application of the rule for equality having the form

$$\frac{t=0,\ \Gamma\to\Delta,\ [A|^{\alpha}_{\rho'}]}{t=0,\ \Gamma\to\Delta,\ [A|^{\alpha}_{\rho}]}\to =,$$

where $\rho \mathrel{\underset{\circ}{=}} 0, t$; $\rho^1 \mathrel{\underset{\circ}{=}} 0, t$ and $\rho \not\mathrel{\underset{\circ}{=}} \rho^1$, by the following figure:

$$\left.\begin{array}{c} t=0,\ \Gamma\to\Delta,\ [A|^{\alpha}_{\rho'}] \\ \cdots \\ \hline t=0,\ \Gamma\to\Delta,\ [A|^{\alpha}_{\rho}] \end{array}\right\} D,$$

where D is an axiomatic deduction of the term ρ from ρ'.

The formula $t = 0$ is not an auxiliary formula of any application of the rule for equality in the deduction V_1^* thus constructed. Eliminating the formula $t = 0$ from V_1^*, we obtain a weeded succedent deduction of the sequent $\Gamma \to \Delta$ in $\mathbf{R}'$ which, according to Theorem 2, can be reconstructed into a deduction in $\mathbf{J}$.

Case 2. $l > 0$, i.e. $t \Rightarrow 0^{[l]}$ in $\mathbf{S}$. We then reconstruct the right branch of the deduction V. As in Case 1, we reconstruct the deduction V_2 into the succedent deduction V_2' in the calculus $\mathbf{R}'$, which we reconstruct in turn into the deduction V_2^* which does not contain applications of rules for equality whose auxiliary formula is the formula $t = (Pd(t))'$. Namely, we replace an arbitrary application of the rule for equality having the form

$$\frac{t=(Pd\,(t))',\ \Gamma\to\Delta,\ [A|^{\alpha}_{t}]}{t=(Pd\,(t))',\ \Gamma\to\Delta,\ [A|^{\alpha}_{(Pd(t))'}]}\to =$$

by the figure

$$\begin{array}{cl} \left.\begin{array}{c} t=(Pd\,(t))',\ \Gamma\to\Delta,\ [A|^{\alpha}_{t}] \\ \cdots \\ \hline t=(Pd\,(t))',\ \Gamma\to\Delta,\ [A|^{\alpha}_{(0^{[l-1]})'}] \end{array}\right\} D_1 & \\ \hline t=(Pd\,(t))',\ \Gamma\to\Delta,\ [A|^{\alpha}_{(Pd(0^{[l]}))'}] & 1.1 \\ \left.\begin{array}{c} \cdots \\ \hline t=(Pd\,(t))',\ \Gamma\to\Delta,\ [A|^{\alpha}_{(Pd(t))'}] \end{array}\right\} D_2, & \end{array}$$

where D_1 is an axiomatic deduction of the term $0^{[l]}$ from the term t and D_2 is an axiomatic deduction of t from $0^{[l]}$.

We note that the application of the axiomatic rule for equality 1.1 is possible in the deduction constructed since $l > 0$ and $0^{[l]} \mathrel{\underset{\circ}{=}} (0^{[l-1]})'$.

Application of rules for equality having the form

$$\frac{t=(Pd\,(t))',\ \Gamma\to\Delta,\ [A|^{\alpha}_{(Pd(t))'}]}{t=(Pd\,(t))',\ \Gamma\to\Delta,\ [A|^{\alpha}_{t}]}\to =$$

is replaced by a figure analogous to the preceding (here we use rule 2.1 instead of 1.1).

Eliminating the formula $t = (Pd(t))'$ from the deduction V_2^*, we obtain the weeded succedent deduction of the sequent $\Gamma \to \Delta$ in $\mathbf{R}'$ which, according to Theorem 2, can be reconstructed into a deduction in **J**. This completes the proof of the lemma.

We shall call an application of rule (4) *normal* if none of the side formulas of this application is a successor of the principal formula of the antecedent axiom. If only no right side formula of an application of rule (4) is a successor of the principal formula of an antecedent axiom, we shall call such an application of rule (4) *seminormal*.(¹) We shall call an application of rule (4) *degenerate* if a right side formula of this application is a successor of the principal formula of an antecedent axiom and for a special term t of such an application there exists a $k \geqslant 0$ such that $\vdash t \Rightarrow 0^{[k]}$ in **S**.

THEOREM 3. *Any deduction in the calculus* **R** *can be reconstructed into a regular deduction in the same calculus with the same sequent in which all applications of rule* (4) *are either normal or seminormal.*

PROOF. The initial deduction V can contain applications of rule (4) whose right side formula is a successor of the principal formula of the antecedent axiom and for a special term t of this application and for arbitrary $k \geqslant 0$ the object $t \Rightarrow 0^{[k]}$ is not deducible in the calculus **S**. To eliminate such applications of rule (4) in V we reconstruct the initial deduction, in accordance with Theorem 2, into the deduction V' in the calculus $\mathbf{R}_{\text{init}}$. Using Theorem 2 and Lemmas 5 and 6, we reconstruct V' into a weeded regular deduction V^* in **R**, where in V^* all applications of rule (4) are either normal, or seminormal, or degenerate. We shall prove the possibility of reconstructing V^* into a deduction not containing degenerate applications of rule (4) by induction on the number n of applications of rule (4) in V^*. The initial step of the induction is trivial.

Induction step: $n > 0$. We consider two cases.

Case 1. Suppose the deduction V^* contains no degenerate applications of rule (4). Then V^* is the required deduction.

Case 2. Suppose V^* contains m degenerate applications of rule (4), where $m \leqslant n$. We shall construct a degenerate application of rule (4). We denote the deduction of the conclusion of this degenerate application of rule (4) by V_1. It follows from the definition of a degenerate application of rule (4) that for a special term t of this application there exists a $k \geqslant 0$ such that $\vdash t \Rightarrow 0^{[k]}$ in **S**. We consider two cases.

a) $k = 0$; in V_1 we then lift up the degenerate application of rule (4) in accordance with Lemma 12. We obtain a deduction V_1' (with the same last sequent)

(¹) In [4] such an application of rule (4) is called *special*.

in which, according to Lemma 12, there are no further applications of rule (4) above the left premise of an arbitrary application of a successor of the selected application of rule (4), i.e. the deduction of a left premise of an arbitrary successor is a deduction in **J**.

Moreover, it follows from condition 3) of Lemma 12 that in V_1' the number of applications of rule (4) which are not successors of the selected application of rule (4) and appear below applications of successors of the selected application of rule (4) is not greater than the number of applications of rule (4) in V_1. According to Lemma 13, an arbitrary deduction whose last step is the application of a successor of the selected application of rule (4) can be reconstructed into a deduction in **J** with the same last sequent. Thus V_1 is reconstructed into a deduction V_1^* containing only those applications of rule (4) which were below the successors of the selected application of rule (4) in V_1', and not containing a degenerate application of rule (4) selected in the reconstruction of the deduction, i.e. the number of applications of rule (4) in V_1^* has been decreased in comparison with the number of applications of rule (4) in V_1. We denote by V^{**} the deduction obtained from V^* after replacing its subdeduction V_1 by V_1^*. The number n of applications of rule (4) in V^{**} has been decreased, although, as follows from the reconstruction of V_1' into V_1^* (see the proof of Lemma 13), the number m of degenerate applications of rule (4) could increase;(2) however, $m \leqslant n$. Since the generic number n of applications of rule (4) in V^{**} has been decreased, the induction assumption is applicable to V^{**}.

b) $k > 0$. This case is analogous to case a). This completes the proof of the theorem.

We shall call a successor $\mathfrak{B}$ of some formula $\mathfrak{A}$ not having successors in a given deduction in the calculus **R** a *fundamental successor* of $\mathfrak{A}$. It follows from the definition of a fundamental successor that it either occurs in the last sequent of the deduction or is a side formula of some application of rule (4) in the given deduction in **R**.

We shall call an application of the rule for equality in a deduction in **R** *reduced* if all successors of the auxiliary formula $\mathfrak{A}$ of this application and the formula $\mathfrak{A}$ itself are subformulas, possibly improper,(3) of the fundamental successor of the formula $\mathfrak{A}$.

We shall call an arbitrary regular deduction in **R** in which all applications of

---

(2) We note that the deduction V_1' can be reconstructed into the deduction V_1^* by Lemma 13, using furthermore Lemma 5; therefore there is no nondegenerate application of rule (4) in V^{**} whose side formula is a successor of the principal formula of an antecedent axiom.

(3) Formula $\mathfrak{A}$ as well as its successors are proper subformulas of the fundamental successor of formula $\mathfrak{A}$ if and only if the fundamental successor of formula $\mathfrak{A}$ appears in the last sequent of the deduction and formula $\mathfrak{A}$ transforms into it by logical rules.

rules for equality are reduced a *reduced deduction.* In an arbitrary reduced deduction, only formulas whose predecessors are not the auxiliary formula of any application of the rule for equality can be principal formulas of the application of rules for equality and axiomatic rules for equality.

LEMMA 14. *An arbitrary deduction in the calculus* **R** *can be reconstructed into a reduced deduction in* **R**, *with the same last sequent and the same fundamental successors of the principal formulas of the axioms as in the initial deduction.*

PROOF. We reconstruct the given deduction into a regular deduction V. The proof is by induction on the number k of nonreduced applications of the rule for equality in V.

The initial step of the induction is trivial.

Induction step: $k > 0$. We consider the lowest nonreduced application of the rule for equality in V. We write out the following subdeduction V_1 of V, which is a deduction in the calculus **J** and includes the application of the rule under consideration:

$$V_1\left\{\begin{array}{l} \quad\quad\quad\quad\cdots \\ \hline p=q,\ \Gamma,\ \Sigma,\ r=s\to\Delta_1,\ [A|^{\alpha}_{q}] \\ \hline V_2\left\{\begin{array}{l} p=q,\ \Gamma,\ \Sigma,\ r=s\to\Delta_1,\ [A|^{\alpha}_{p}] \\ \quad\quad\quad\quad\cdots \\ \hline \quad\quad\quad r=s,\ \Gamma\to\Delta \end{array}\right. \end{array}\right. \to =$$

where $p = q$ is an auxiliary formula of the nonreduced application of the rule for equality; $r = s$ is a successor of $p = q$ which is a subformula of the fundamental successor of $p = q$; V_2 is a deduction consisting of application of the rules for equality and rules (1′), (2′), and including rules carrying the term p into the term r and the term q into the term s; and Σ is a list of formulas which appear due to doubling. We replace V_1 by the following deduction, V_1':

$$V_1'\left\{\begin{array}{l} \quad\quad\quad\quad\cdots \\ \hline V_2''\left\{\begin{array}{l} p=q,\ \Gamma,\ \Sigma,\ r=s\to\Delta_1,\ [A|^{\alpha}_{q}] \\ \quad\quad\quad\quad\cdots \\ \hline p=q,\ \Gamma,\ \Sigma,\ r=s\to\Delta_1,\ [A|^{\alpha}_{s}] \end{array}\right. \\ \hline V_2'\left\{\begin{array}{l} p=q,\ \Gamma,\ \Sigma,\ r=s\to\Delta_1,\ [A|^{\alpha}_{r}] \\ \quad\quad\quad\quad\cdots \\ \hline p=q,\ \Gamma,\ \Sigma,\ r=s\to\Delta_1,\ [A|^{\alpha}_{p}] \end{array}\right. \\ V_2\left\{\begin{array}{l} \quad\quad\quad\quad\cdots \\ \hline \quad\quad\quad r=s,\ \Gamma\to\Delta \end{array}\right. \end{array}\right. \to =$$

where the auxiliary formula of the explicitly indicated application of the rule for equality is the formula $r = s$; V_2' is the deduction consisting of applications of rules which carry the term r into the term p and are constructed according to deduction V_2; V_2'' is the deduction consisting of applications of rules which carry the term q

into the term s and are constructed according to deduction V_2.

Replacing the subdeduction V_1 of V by the deduction V'_1, we obtain the deduction V' in which the number of nonreduced applications of rules for equality is one less than in V, i.e. the induction hypothesis is applicable to V'. It is clear from the reconstructions carried out above that V'_1 had no applications of rules whose principal formula was not subject to a transformation in V, i.e. the fundamental successors of principal formulas of axioms in the reconstructed deduction will be the same as in the initial deduction.

The case when the principal formula of a nonreduced application of the rule for equality occurs in the antecedent is considered analogously.

REMARK 3. An arbitrary weeded reduced deduction in the calculus **R** does not contain applications of rules for equality and rules (1′) and (2′) whose principal formula is not the successor of a principal formula of an axiom.

We note that Lemma 14 and Remark 3 generalize trivially to the case of an arbitrary calculus, corresponding to an axiomatically given theory with equality (with corresponding definition of a reduced deduction).

LEMMA 15. *Let* V *be a deduction in the calculus* **R** *whose last step consists of a seminormal application of rule* (4) *and above whose implications appear reduced deductions in the calculus* **J**. *Then one can construct a deduction* V' *with the same last sequent which either is a deduction in* **J** *or the last step of the deduction consists in a normal application of rule* (4) *and above the premises of this application there appear deductions in* **J**; *and, moreover, formulas which are not subject to change in* V *will also not change in* V'.

PROOF. The end of the given deduction V has the form

$$\frac{V_1\left\{\frac{\cdots}{t=0,\ \Gamma\to\Delta};\quad V_2\left\{\frac{\cdots}{t=(Pd\,(t))',\ \Gamma\to\Delta}\right.\right.}{\Gamma\to\Delta}\,(4).$$

Since the indicated application of rule (4) is seminormal, the principal formula of the axiom of the deduction V_2 is not the successor of the selected occurrence of the formula $t=(Pd(t))'$. We consider two cases.

Case 1. Suppose that in V_2 there are no applications of the rule for equality whose auxiliary formula is a predecessor of the selected occurrence of $t=(Pd(t))'$. Then eliminating the formula $t=(Pd(t))'$ (all predecessors of this formula coincide with the formula itself since V_2 is reduced) from all sequents of V_2, we obtain the desired deduction V' in **J**.

Case 2. Suppose that in V_2 there are applications of the rule for equality whose auxiliary formula is a predecessor of the selected occurrence of the formula $t=(Pd(t))'$. A successor of the principal formula of an axiom of V_2 is a formula from the list

Γ, Δ. We shall show that the deduction V_1 of the left premise of application of rule (4) in V can be reconstructed into the deduction V_1' (with the same last sequent) in **J** in which the principal formula of an axiom does not have the same fundamental successor as the principal formula of an axiom in V_2. We then obtain the desired deduction V' from V by replacing in it the deduction of the left premise V_1 by the deduction V_1'.

We construct the deduction V_1' whose last sequent has the form $t = 0,\ \Gamma \to \Delta$ by adding the formula $t = 0$ to all sequents of V_2, eliminating the formula $t = (Pd(t))'$ and replacing in V_2 every application of the rule for equality with the auxiliary formula $t = (Pd(t))'$ by some configuration. In constructing the configuration to replace applications of the rule for equality by the auxiliary formula $t = (Pd(t))'$, we use the fact that in the deduction V_1 of the left premise of a seminormal application of rule (4) the side formula $t = 0$ is a successor of the principal formula of an antecedent axiom, i.e. there exists an $l \geqslant 0$ such that $t = 0$ is the successor of a formula of the form $r^{[l+1]} = 0^{[l]}$ or $0^{[l]} = r^{[l+1]}$. In other words, having in the antecedent of some sequent a list of the formulas $t = 0,\ \Gamma$ and having some formula $[A|_t^\alpha]$, we can construct $k > 0$ terms q and a deduction tree in which the formula $[A|_s^\alpha]$, where $s \mathrel{\overline{\underline{\circ}}} q^{[k]}$, is a predecessor (or successor) of the formula $[A|_t^\alpha]$. The assertion is clear if 1) $l = 0$ and the formula $t = 0$ is a successor of the formula $r^{[l+1]} = 0^{[l]}$, or 2) $l > 0$. It remains to consider the case when $l = 0$ and the formula $t = 0$ is a successor of the formula $0 = r'$, i.e. the term t in the deduction V_1 goes over into the term 0. Then the term 0 must go over into the term r', i.e. t must be carried over into r'.

Suppose the application under consideration of the rule for equality has the form

$$\frac{t = (Pd\,(t))',\ \Gamma \to \Delta_1,\ [A\,|^\alpha_{(Pd(t))'}]}{t = (Pd\,(t))',\ \Gamma \to \Delta_1,\ [A\,|^\alpha_t]} \to =$$

We replace this application by the following figure:

$$\begin{array}{l} D_1 \left\{ \begin{array}{c} t = 0,\ \Gamma \to \Delta_1,\ [A\,|^\alpha_{(Pd(t))'}] \\ \cdots \\ \hline t = 0,\ \Gamma \to \Delta_1,\ [A\,|^\alpha_{(Pd(q^{[k]}))'}] \end{array} \right. \\ \hline D_2 \left\{ \begin{array}{c} t = 0,\ \Gamma \to \Delta_1,\ [A\,|^\alpha_{q^{[k]}}] \\ \cdots \\ \hline t = 0,\ \Gamma \to \Delta_1,\ [A\,|^\alpha_t] \end{array} \right. \end{array} \; 2.1 \quad ,$$

where D_1 consists of the application of rules which carry the term t into the term $q^{[k]}$ and constructed according to the deduction V_1; D_2 consists of the application of rules which carry $q^{[k]}$ into t and are constructed according to V_1. We note that the term $Pd(q^{[k]})$ is replaced in the application of rule 2.1; this is possible since $k > 0$. If the application of the rule for equality has the form

$$\frac{t = (Pd\,(t))',\ \Gamma \to \Delta_1,\ [A\,|^{\alpha}_{t}]}{t = (Pd\,(t))',\ \Gamma \to \Delta_1,\ [A\,|^{\alpha}_{(Pd(t))'}]} \to =,$$

the figure which replaces this application of the rule is constructed exactly analogously to the preceding figure, where rule 1.1 is used instead of rule 2.1.

The case of an antecedent application of the rule for equality is considered analogously to the cases of the succedent application of the rule for equality.

This completes the proof of the lemma.

THEOREM 4. *An arbitrary deduction in the calculus* **R** *can be reconstructed into a regular deduction in* **R** *and with the same last sequent, containing only normal applications of the rule* (4).

PROOF. We reconstruct the given deduction V, according to Theorem 3, into a regular deduction V' containing only normal and seminormal applications of rule (4). To reconstruct V' into the desired deduction V^* we carry out an induction on the number n of seminormal applications of rule (4) in V'.

The initial step of the induction is trivial.

Induction step: $n > 0$. By Lemma 11 we lift up the highest left seminormal application of rule (4). We obtain the deduction V'' in which there are no further applications(4) of rule (4) above an arbitrary successor of the selected seminormal application of rule (4). According to Lemmas 14 and 15, all the subdeductions of V'' whose last steps consist in the application of rule (4), being the successors of the selected application of rule (4), can be reconstructed either into deductions in the calculus **J** or into deductions whose last steps are normal applications of rule (4). It follows from Lemmas 11 and 15 that no new seminormal applications of rule (4) will appear in the reconstructed deduction. Thus the induction parameter decreases. This completes the proof of the theorem.

We shall call an arbitrary reduced deduction in the calculus **R** in which all the applications of rule (4) are normal a *normal deduction*. The following theorem follows in an obvious way from Theorems 3 and 4, Lemmas 6 and 14, and Remark 3.

THEOREM 5. *An arbitrary deduction in the calculus* **R** *can be reconstructed into a normal deduction in* **R** *with the same last sequent.*

§5. Equivalence of the calculi R and $\mathbf{R}_{\text{fin}}$

Let $\mathbf{R}^*$ be the calculus obtained from the basic calculus $\mathbf{R}_{\text{fin}}$ by replacing the scheme of rule (7) by the following scheme of the rule:

---

(4) Under such a reconstruction of the deduction V' according to Lemma 11 into the deduction V'', the number of seminormal applications of rule (4) which are not successors of the selected application of rule (4) does not increase inasmuch as there are no further seminormal applications of rule (4) above the selected one.

$$\frac{t=0,\ \Gamma_1 \to \Delta_1;\ \ t=(Pd\,(t))',\ \Gamma_2 \to \Delta_2}{\Gamma \to \Delta}, \tag{8}$$

where the notation is the same as in rule (7), i.e. the scheme of rule (8) differs from the scheme of rule (7) by the presence of side formulas of rule (4).

The concepts of weeded deduction, fundamental successor of a formula, reduced application of a rule, and so on, introduced in the preceding sections, generalize in a natural way to **R***. We shall call the formulas introduced explicitly in rule (8) *quasi-side* formulas of rule (8) (in the absence of side formulas of this rule in usual terminology).

We shall call an application of rule (8) *normal* if no quasi-side formula of this application is the successor of the principal formula of an antecedent axiom. We shall call an arbitrary reduced deduction in **R*** in which all applications of rule (8) are normal a *normal deduction* in **R***.

LEMMA 16. *Given a normal weeded deduction V whose last step is an application of rule* (4) *and above whose premise appear normal deductions in the calculus* **R*** *which do not contain applications of logical rules, one can construct a normal deduction with the same last sequent in* **R***, *and, moreover, there are no applications of logical rules in the constructed deduction.*

PROOF. The end of the given deduction has the form

$$V\left\{\frac{V_1\left\{\dfrac{\cdots}{t=0,\ \Gamma \to \Delta;}\right. \quad V_2\left\{\dfrac{\cdots}{t=(Pd\,(t))',\ \Gamma \to \Delta}\right.}{\Gamma \to \Delta}\,(4).\right.$$

Since the deduction V is weeded and the application of rule (4) is normal, in each of the deductions V_1 and V_2 there must be applications of the rule for equality whose auxiliary formulas are predecessors of side formulas of rule (4). We consider the lowest applications of such rules and denote them by $R_1, \cdots, R_k$. We denote by l_i $(i = 1, \cdots, k)$ the number of applications of rule (8) appearing between the application R_i $(i = 1, \cdots, k)$ and the last sequent of V. We prove the lemma by induction on $(l_1 + \cdots + l_k)$.

Initial step: $(l_1 + \cdots + l_k) = 0$. Then $k = 2$. We consider the case when the principal formulas of applications R_1 and R_2 have the same formula, occurring, for definiteness, in the succedent of the last sequent, as fundamental successor in V. Then V has the form

$$V\left\{\frac{V_1\left\{\begin{array}{l} V_1'\left\{\dfrac{\cdots}{t=0,\ \Gamma,\ \Pi_1 \to \Sigma_1,\ [B|^{\alpha}_{\rho_1'}]}\right. \\ \hline D_1\left\{\begin{array}{c} t=0,\ \Gamma,\ \Pi_1 \to \Sigma,\ [B|^{\alpha}_{\rho_1}],\ \theta_1 \\ \cdots \\ \hline t=0,\ \Gamma \to \Delta,\ A \end{array}\right. \end{array}\right. R_1 \quad \left.\begin{array}{l} V_2'\left\{\dfrac{\cdots}{t=(Pd\,(t))',\ \Gamma,\ \Pi_2 \to \Sigma_2,\ [C|^{\beta}_{\rho_2'}]}\right. \\ \hline D_2\left\{\begin{array}{c} t=(Pd\,(t))',\ \Gamma,\ \Pi_2 \to \Sigma,\ [C|^{\beta}_{\rho_2}],\ \theta_2 \\ \cdots \\ \hline t=(Pd\,(t))',\ \Gamma \to \Delta,\ A \end{array}\right. \end{array} R_2\right\} V_2}{\Gamma \to \Delta,\ A}\,(4),\right.$$

where $\Sigma \overset{\circ}{=} \Delta,\ A$; $\Sigma_1 \overset{\circ}{=} \Sigma,\ [B|^{\alpha}_{\rho_1}], \Theta_1$; $\Sigma_2 \overset{\circ}{=} \Sigma,\ [C|^{\alpha}_{\rho_2}], \Theta_2$; $\Pi_1, \Pi_2, \Theta_1, \Theta_2$ are lists of formulas appearing in the implications of applications of rules for equality and the rules (1*), (2*) in the deduction V; $\rho_1 \overset{\circ}{=} 0, t$; $\rho_1' \overset{\circ}{=} 0, t$; $\rho_1 \overset{\circ}{\neq} \rho_1'$; $\rho_2 \overset{\circ}{=} t, (Pd(t))'$; $\rho_2' \overset{\circ}{=} t, (Pd(t))'$, $\rho_2 \overset{\circ}{\neq} \rho_2'$, and A is a formula which is a fundamental successor of the formulas $[B|^{\alpha}_{\rho_1}]$ and $[C|^{\beta}_{\rho_2}]$.

Using doubling of antecedent and succedent principal formulas in the rules for equality and in rules (1*) and (2*), we reconstruct V into a deduction in **R*** in the following way:

$$
\begin{array}{l}
V_1''\left\{\dfrac{\cdots}{t=0,\ \Gamma,\ \Pi_1,\ \Pi_2\to\Sigma_1,\ \Sigma_3,\ [B|^{\alpha}_{\rho_1'}];}\right. \quad V_2''\left\{\dfrac{\cdots}{t=(Pd\,(t))',\ \Gamma,\ \Pi_1,\ \Pi_2\to\Sigma_1,\ \Sigma_3,\ [C|^{\beta}_{\rho_2'}]}\right. \\
\hline
D_2'\left\{\begin{array}{c} \Gamma,\ \Pi_1,\ \Pi_2\to\Delta,\ A,\ [B|^{\alpha}_{\rho_1}],\ \Theta_1,\ [C|^{\beta}_{\rho_2}],\ \Theta_2 \qquad\qquad (8)\\ \cdots \\ \hline \end{array}\right. \\
D_1\left\{\begin{array}{c} \Gamma,\ \Pi_1\to\Delta,\ A,\ [B|^{\alpha}_{\rho_1}],\ \Theta_1 \\ \cdots \\ \hline \Gamma\to\Delta,\ A \end{array}\right. ,
\end{array}
$$

where the notation is the same as in the deduction V and $\Sigma_3 \overset{\circ}{=} [C|^{\beta}_{\rho_2}], \Theta_2$.

Since R_i $(i = 1, 2)$ is the lowest application of the rule for equality with auxiliary formula which is a side formula of the application of rule (4), these side formulas do not change in the transitions D_1 and D_2. The deductions V_1'', V_2'' and D_2' are obtained respectively from V_1', V_2' and D_2 by adding certain formulas. Since succedent and antecedent principal formulas in the rules for equality and in rules (1*) and (2*) are doubled, all other cases of the initial induction step are considered in precisely the same way as the case we just considered.

Induction step: $l_1 + \cdots + l_k > 0$. Then there is in V_1 or V_2 at least one application of rule (8) which appears lower than the applications $R_1, \cdots, R_k$. We consider the lowest left application of this sort. Suppose the deduction V has the form

$$
V\left\{\begin{array}{l}
V_1\left\{\begin{array}{l}
V_1'\left\{\dfrac{\cdots}{r=0,\ t=0,\ \Gamma_1,\ \Pi_1\to\Delta_1,\ \Sigma_1;}\right. \quad V_1''\left\{\dfrac{\cdots}{r=(Pd\,(r))',\ t=0,\ \Gamma_2,\ \Pi_2\to\Delta_2,\ \Sigma_2}\right. \\
\hline
D\left\{\begin{array}{c} t=0,\ \Gamma,\ \Pi\to\Delta,\ \Sigma \qquad\qquad (8) \\ \cdots \\ \hline t=0,\ \Gamma\to\Delta; \end{array}\right.
\end{array}\right. \quad V_2\left\{\dfrac{\cdots}{t=(Pd\,(t))',\ \Gamma\to\Delta}\right. \\
\hline
\qquad\qquad \Gamma\to\Delta \qquad\qquad (4),
\end{array}\right.
$$

where Π and Σ are lists of formulas appearing in the premises of applications of the rules for equality and rules (1*) and (2*); $\Gamma_1, \Pi_1, \Delta_1, \Sigma_1, \Gamma_2, \Pi_2, \Delta_2$ and Σ_2 are formulas obtained as a result of the concrete realization of the scheme of rule (8); D does not contain applications of rule (8), since it is considered lower than the application of rule (8) in V and, furthermore, D contains no applications of rules for equality with auxiliary formula $t = 0$. We reconstruct V as follows:

$$\cfrac{\cfrac{V_1'\left\{\cfrac{\cdots}{t=0,\ \Omega_1\to\Theta_1;}\right. \quad V_2'\left\{\cfrac{\cdots}{t=(Pd(t))',\ \Omega_1\to\Theta_1}\right.}{r=0,\ \Gamma_1,\ \Pi_1\to\Delta_1,\ \Sigma_1}\,(4);\quad \cfrac{V_1''\left\{\cfrac{\cdots}{t=0,\ \Omega_2\to\Theta_2;}\right. \quad V_2''\left\{\cfrac{\cdots}{t=(Pd(t))',\ \Omega_2\to\Theta_2}\right.}{r=(Pd(r))',\ \Gamma_2,\ \Pi_2\to\Delta_2,\ \Sigma_2}\,(4)}{D\left\{\begin{array}{c}\Gamma,\ \Pi\to\Delta,\ \Sigma\\ \cdots\\ \hline \Gamma\to\Delta\end{array}\right.}\,(8),$$

where $\Omega_1 \rightleftharpoons r = 0, \Gamma_1, \Pi_1$; $\Theta_1 \rightleftharpoons \Delta_1, \Sigma_1$; $\Omega_2 \rightleftharpoons r = (Pd(r))', \Gamma_2, \Pi_2$; and $\Theta_2 \rightleftharpoons \Delta_2, \Sigma_2$.

Deductions V_2' and V_2'' are obtained from V_2 by adding corresponding formulas. Since in the reconstructed deduction the induction parameter $(l_1 + \cdots + l_k)$ decreased in the deductions of premises of explicitly selected applications of rule (4), we have, according to the induction assumption, that each of these deductions can be reconstructed into a deduction in $\mathbf{R}^*$, i.e. the deduction V will also be reconstructed into a deduction in $\mathbf{R}^*$. It is clear from the reconstructions that the reconstructed deduction in $\mathbf{R}^*$ will be normal, and, furthermore, it will not contain applications of logical rules.

THEOREM 6. *Given a normal weeded deduction in* $\mathbf{R}$, *one can construct a normal deduction in* $\mathbf{R}^*$ *with the same last sequent.*

PROOF. In the given deduction V in $\mathbf{R}$ all succedent applications of rules for equality and rules (1′) and (2′) are replaced by the corresponding applications of rules in $\mathbf{R}^*$. Further, we apply induction on the number n of applications of rule (4) in V. We use Lemma 16 in the induction step to reconstruct the deduction.

Instead of the scheme of rule (8) we introduce the scheme of the rule

$$\frac{\Gamma_1 \to \Delta_1;\quad t = (Pd(t))',\ \Gamma_2 \to \Delta_2}{\Gamma \to \Delta}, \tag{9}$$

where the notation is the same as in the scheme of rule (7). The scheme of rule (9) differs from the scheme of rule (8) by the presence of only one quasi-side formula.

We denote by $\mathbf{R}^*_{\mathrm{pr}}$ the calculus obtained from $\mathbf{R}^*$ on replacing the scheme of rule (8) by the scheme of rule (9). A normal application of rule (9) as well as a normal deduction in $\mathbf{R}^*_{\mathrm{pr}}$ are defined in the usual way.

LEMMA 17. *Given a weeded normal deduction* V *whose last step consists in an application of rule* (8) *and above whose premises there appear normal deductions in the calculus* $\mathbf{R}^*_{\mathrm{pr}}$ *which do not contain applications of logical rules, one can construct a normal deduction with the same last sequent in* $\mathbf{R}^*_{\mathrm{pr}}$, *and, furthermore, there will be no applications of logical rules in the constructed deduction.*

PROOF. The end of the given deduction V has the form

$$V\left\{\dfrac{V_1\left\{\dfrac{\cdots}{t=0,\ \Gamma_1\to\Delta_1;}\right. \quad V_2\left\{\dfrac{\cdots}{t=(Pd\,(t))',\ \Gamma_2\to\Delta_2}\right.}{\Gamma\to\Delta}\right. (8),$$

where Γ is a sublist of the list Γ_i $(i = 1, 2)$ and Δ is a sublist of Δ_i $(i = 1, 2)$.

The lemma is proved by induction on the number k of applications of the rules for equality whose auxiliary formula is a predecessor of the selected occurrence of the quasi-side formula $t = 0$ in the deduction V_1.

Initial step: $k = 0$. Since the application of rule (8) is normal and $k = 0$, after eliminating from V_1 the predecessors of the selected occurrence of the formula $t = 0$ no application of the postulate in V_1 is disturbed. The deduction V will be reconstructed into the desired deduction.

Induction step: $k > 0$. We consider the lowest left application of the rule for equality with auxiliary formula $t = 0$ in the deduction V_1. We denote this application by R_1. Suppose V_1 has the form

$$V_1\left\{\begin{array}{l} V_1'\left\{\dfrac{\cdots}{t=0,\ \Gamma_1,\ \Pi\to\Delta_1\ \Theta,\ [A|^{\alpha}_{\rho}],\ [A|^{\alpha}_{\rho'}]}\right. \\ \hline D\left\{\dfrac{t=0,\ \Gamma_1,\ \Pi\to\Delta_1,\ \Theta,\ [A|^{\alpha}_{\rho}]\quad \cdots}{t=0,\ \Gamma_1\to\Delta_1}\right. \end{array}\right. R_1,$$

where $\rho \mathrel{\underline{\circ}} 0, t$; $\rho' \mathrel{\underline{\circ}} 0, t$; $\rho \not\mathrel{\underline{\circ}} \rho'$; and Π, Θ and $[A|^{\alpha}_{\rho}]$ are formulas appearing in premises of applications of rules of the subdeduction D, where there may be applications of rule (9) in D. The principal formula of R_1 is succedent (the case when the principal formula is antecedent is considered precisely analogously).

We shall show that application R_1 can be replaced by some figure in the calculus $\mathbf{R}^*_{\mathrm{pr}}$, since the conclusion of the application R_1 contains all formulas of the list Γ, Δ. Namely, we reconstruct V_1 as follows:

$$V_1''\left\{\begin{array}{l} V_1'\left\{\dfrac{\cdots}{t=0,\ \Gamma_1,\ \Pi\to\Delta_1,\ \Theta,\ [A|^{\alpha}_{\rho}],\ [A|^{\alpha}_{\rho'}];}\right. \quad V_2'\left\{\dfrac{\cdots}{t=(Pd\,(t))',\ t=0,\ \Gamma_3\to\Delta_3}\right. \\ \hline D\left\{\dfrac{t=0,\ \Gamma_1,\ \Pi\to\Delta_1,\ \Theta,\ [A|^{\alpha}_{\rho}]\quad \cdots}{t=0,\ \Gamma_1\to\Delta_1}\right. \end{array}\right. (9)$$

where Γ_3 is a list of formulas containing all formulas from Γ_2, Γ_1 and Π, and Δ_3

is a list of formulas containing all formulas from Δ_2, Δ_1, Θ and $[A|^{\alpha}_{\rho}]$.

The deduction V'_2 is obtained from V_2 by adding corresponding formulas.

The induction parameter k decreases by one in the deduction V''_1, and therefore, applying the induction assumption, we can reconstruct V''_1 into a deduction in $\mathbf{R}^*_{\mathrm{pr}}$. By the same token we reconstruct V into the deduction sought, and furthermore it is clear from the reconstructions that all the applications of rule (9) are normal. This completes the proof of the lemma.

THEOREM 7. *Given a normal weeded deduction in* $\mathbf{R}^*$, *one can construct a normal deduction in* $\mathbf{R}^*_{\mathrm{pr}}$ *with the same last sequent.*

This theorem is proved by induction on the number n of applications of rule (8) in the given deduction. Lemma 17 is used in the induction step.

THEOREM 8. *Given a normal weeded deduction in* $\mathbf{R}^*_{\mathrm{pr}}$, *a deduction in* $\mathbf{R}_{\mathrm{fin}}$ *can be constructed with the same last sequent.*

The theorem is proved analogously to Theorem 7 using induction on the number of applications of rule (9) in the initial deduction. An assertion analogous to Lemma 17 is used at the induction step.

THEOREM 9. *The calculi* $\mathbf{R}$ *and* $\mathbf{R}_{\mathrm{fin}}$ *are equivalent.*

PROOF. According to Theorem 5, an arbitrary deduction in $\mathbf{R}$ can be reconstructed into a normal deduction in $\mathbf{R}$. Carrying out a preliminary weeding, the constructed deduction can be reconstructed, according to Theorems 6, 7 and 8, into a deduction in $\mathbf{R}_{\mathrm{fin}}$.

To reconstruct a deduction V in $\mathbf{R}_{\mathrm{fin}}$ into a deduction in $\mathbf{R}$, we replace each application of rule (7) in V by an application of rule (4) and applications of the rules for equality in every branch. We denote the deduction obtained by V'. We reconstruct V' into a regular weeded deduction V''. No branch of V'' can contain two distinct succedent applications of the rules for equality or the rules (1*) or (2*) whose principal formulas have different formulas as their successors, inasmuch as V'' is weeded. Therefore the formulas, being principal formulas in the application of succedent rules for equality or rules (1*) or (2*) due to doubling, can be eliminated, i.e. by replacing all applications of the rules of $\mathbf{R}_{\mathrm{fin}}$ by applications of the corresponding rules of $\mathbf{R}$, we reconstruct V'' into a deduction in $\mathbf{R}$. This completes the proof of the theorem.

Theorems 2 and 9 imply the following theorem.

THEOREM 10. *The calculi* $\mathbf{R}_{\mathrm{init}}$ *and* $\mathbf{R}_{\mathrm{fin}}$ *are equivalent.*

An obvious consequence of Theorem 10 is the consistency of Raphael M. Robinson's arithmetic system.

REMARK 4. Instead of $\mathbf{R}_{\text{fin}}$, one can consider either of the two calculi obtained from $\mathbf{R}_{\text{fin}}$ by eliminating the axiom scheme (5) or the scheme (6). Using the method applied in proving Theorem 9 and the idea of doubling principal formulas in antecedent rules for equality and antecedent axiomatic rules for equality, one can prove the equivalence of either of the indicated calculi with $\mathbf{R}_{\text{init}}$. We remark, however, that it is more convenient to use $\mathbf{R}_{\text{fin}}$ in looking for the deduction.

BIBLIOGRAPHY

1. A. Ju. Pljuškevičene, *Specialization of the use of axioms for deduction search in axiomatic theories with equality*, Zap. Naučn. Sem. Leningrad. Otdel. Mat. Inst. Steklov. (LOMI) **20** (1971), 175–185 = J. Soviet Math. **1** (1973), 110–116.

2. S. C. Kleene, *Introduction to metamathematics*, Van Nostrand, Princeton, N. J., 1952; Russian transl., IL, Moscow, 1957. MR **14**, 525; **19**, 2.

3. R. A. Pljuškevičus, *A sequential variant of constructive logic calculi for normal formulas not containing structural rules*, Trudy Mat. Inst. Steklov. **98** (1968), 155–202 = Proc. Steklov Inst. Math. **98** (1968), 175–229. MR **38** #4282; **43** #4620.

4. A. Ju. Pljuškevičene, *Elimination of cut-type rules from the Robinson and Presburger axiomatic systems*, Zap. Naučn. Sem. Leningrad. Otdel. Mat. Inst. Steklov. (LOMI) **20** (1971), 186–199 = J. Soviet Math. **1** (1973), 117–125. MR **45** #4962.

5. M. G. Rogava, *Sequential variants of applied predicate calculi without structural deductive rules*, Trudy Mat. Inst. Steklov. **121** (1972), 136–164 = Proc. Steklov Inst. Math. **121** (1972),

6. S. Kanger, *A simplified proof method for elementary logic*, Computer Programming and Formal Systems, North-Holland, Amsterdam, 1963, pp. 87–94; Russian transl. in Mathematical Theory of Logical Inference, "Nauka", Moscow, 1967, pp. 200–209. MR **27** #1350.

7. V. A. Lifšic, *Specialization of the form of deduction in the predicate calculus with equality and function symbols.* I, Trudy Mat. Inst. Steklov. **98** (1968), 5–25 = Proc. Steklov Inst. Math. **98** (1968), 1–24. MR **40** #5415; **43** #4620.

8. V. P. Orevkov, *On Glivenko sequent classes*, Trudy Mat. Inst. Steklov. **98** (1968), 131–154 = Proc. Steklov Inst. Math. **98** (1968), 147–176. MR **39** #2606; **43** #4620.

Translated by L. F. BORON

Trudy Mat. Inst. Steklov.
121 (1972)

Proc. Steklov Inst. Math.
121 (1972)

SEQUENTIAL VARIANTS OF APPLIED PREDICATE CALCULI WITHOUT STRUCTURAL DEDUCTIVE RULES

UDC 51.01: 164

M. G. ROGAVA

ABSTRACT. A result is established concerning specialization of the form of inference in a predicate calculus with equality (the possibility of being restricted to "consequent applications" of rules for equality), which admits numerous applications. As an application of this lemma, a theorem is proved on normalization for a predicate calculus with equality.

Gentzen variants which are cut-free and do not contain other structural rules are constructed for axiomatically characterized mathematical theories prescribed by a finite list, or a denumerable set, of specific axioms and having as apparatus of logical inference the logical tools of the classical first-order predicate calculus with constant function symbols and equality. (Any such mathematical theory is called an applied predicate calculus.)

Bibliography: 22 items.

CONTENTS.

Introduction

G. Gentzen in [1] constructed a sequential variant of the classical first-order predicate calculus without a cut rule (the *LK* calculus). This variant of the predicate calculus proved to be a convenient starting point for the study of a number of important problems in the theory of logical inference and for the construction of various search-for-deduction algorithms. Gentzen's theorem on the normal form of deductions in the predicate calculus is the theoretical basis of many of the algorithms developed today for the machine proof of theorems. It provides the opportunity to construct deduction search algorithms in the predicate calculus which appropriately take into account the concrete peculiarities of the formula tested. Theoretically this theorem provides the opportunity to construct deduction search algorithms in arbitrary finitely axiomatizable theories: the deducibility of a formula in a finitely axiomatizable theory is equivalent

---

AMS (MOS) subject classifications (1970). Primary 02B10; Secondary 68A40.

to the deducibility in the pure predicate calculus of an implication whose premise is a conjunction of axioms and whose conclusion is the formula tested for deducibility.

However, algorithms of the type described which are based directly on Gentzen's theorem for the predicate calculus do not take into account the peculiarities of the axiomatic theory under consideration. It is therefore natural to expect that algorithms based on analogues of Gentzen's theorem for concrete axiomatic theories would prove to be more effective. Until recently theorems of that kind were absent from the literature.

S. Kanger [2], V. A. Matulis [3], [4] and H. B. Curry [5] have constructed sequential variants of the classical first-order predicate calculus, also without a cut rule, which possess, moreover, a number of characteristics particularly vital for the solution of the problem of logical deduction search (reversibility of all rules of deduction, absence of structural rules, and "rigid" conditions for the variation of terms in quantifier rules). S Kanger in [6] found a sequential variant possessing characteristics of the described type also for the classical first-order predicate calculus with equality.

Sequential variants analogous in certain respects are constructed in the present paper for axiomatically characterized mathematical theories which are prescribed by a finite list, or denumerable set, of specific axioms and have as apparatus of logical inference the logical tools of the classical first-order predicate calculus with constant function symbols and equality. (Below such mathematical theories are called applied predicate calculi.) In the sequential variants of applied predicate calculi which are proposed here we manage to replace specific axioms by deductive rules so that systems are obtained which are more suited to searching for a logical deduction than are systems of the types described above. In the case when a specific axiom has the form of an equality of two terms the rules of deduction replacing that specific axiom have a particularly simple form.

To eliminate the cut rule from applied predicate calculi, Gentzen's method must be suppliemented by a number of devices not provided by that method.

On the basis of this paper's results, published without proof in [7], Gentzen variants of calculi of constructive logic were proposed in [8] by R. A. Pljuškevičus, while in [9], under some restrictions on the matrices of the specific axioms, a method was found for replacing cut type rules relative to subformulas of the specific axioms by axiom schemes, and a method is indicated for constructing a sequential variant for an arbitrary axiomatic theory.

When constructing applied predicate calculi, we use two type of object variables: free variables $a_1, a_2, \cdots$ and bound variables $x_1, x_2, \cdots$. The expression $[F|^{u_1,\dots,u_k}_{v_1,\dots,v_k}]$ will denote the result of simultaneous substitution into the word F of the word v_1 in place of each occurrence of the word $u_1, \cdots$, of the word v_k in place of each occurrence of u_k.(¹)

When speaking of a deduction in some calculus, we shall always assume that it can be written in the form of a tree. The height of a deduction is defined as the number of sequents in its longest branch.

Let us call a word $\mathfrak{P}$ a *quasi-term* (*quasi-formula*) if there are a term (respectively, formula) F, free variables $\alpha_1, \cdots, \alpha_p$ $(p \geqslant 0)$ and bound variables $z_1, \cdots, z_p$ not occurring in F such that $\mathfrak{P}$ coincides with $[F|^{\alpha_1,\ldots,\alpha_p}_{z_1,\ldots,z_p}]$.

It is easy to see that any term (any formula) is a quasi-term (a quasi-formula). Quasi-terms (quasi-formulas) differ from terms (from formulas) only in that in them there are occurrences of bound variables which are not bound by quantifiers.

Let C_1 be some set of terms and C_2 some set of quasi-terms. Let us say, following [10], that a term T is *immediately accessible* from C_1 and C_2 if one of the following three conditions is fulfilled:

1) T belongs to C_1.

2) There are a quasi-term t of C_2, bound variables $z_1, \cdots, z_l$ occurring in t, and terms $s_1, \cdots, s_l$ of C_1 such that T coincides with $[t|^{z_1,\ldots,z_l}_{s_1,\ldots,s_l}]$.

3) C_1 is empty and T is the variable a_1.

Let f be some n-place function symbol. We denote by $C(f)$ the quasi-term $f(x_1, \cdots, x_n)$, and denote by $C(f_1, \cdots, f_m)$, where $f_1, \cdots, f_m$ are function symbols, the list of quasi-terms $C(f_1), \cdots, C(f_m)$.

Following [3], [4], let us call a sequent $\Gamma \rightarrow \Delta$ a *prime sequent* if each member of the list Γ is either an elementary formula or a formula beginning with the quantifier $\forall$, while each member of the list Δ is either an elementary formula or a formula beginning with the quantifier $\exists$. We call a sequent *elementary* if all formulas occurring in that sequent are elementary.

Let us consider the closed formula $\forall \alpha_1 \cdots \alpha_k M$, where $\alpha_1, \cdots, \alpha_k$ are arbitrary bound variables and M is a quantifier-free quasi-formula (called the matrix of this formula). Let us call any formula of the form $[M|^{\alpha_1,\ldots,\alpha_k}_{T_1,\ldots,T_k}]$, where $T_1, \cdots, T_k$ are terms, an *instance* of the formula $\forall \alpha_1 \cdots \alpha_k M$, and let us call any formula of the form

$$[\forall \alpha_{i+1} \cdots \alpha_k M |^{\alpha_1,\ldots,\alpha_i}_{T_1,\ldots,T_i}] \ (i > 0)$$

a *quasi-instance* of the formula $\forall \alpha_1 \cdots \alpha_k M$. It is obvious that any instance of $\forall \alpha_1 \cdots \alpha_k M$ is a quantifier-free formula.

The expression $\mathfrak{P} \overline{\overline{\circ}} \mathfrak{Q}$ denotes the graphic equality of the words $\mathfrak{P}$ and $\mathfrak{Q}$.

In this paper the letters t, r, s, u and w, with or without indices, are employed as metamathematical variables for terms; the letters A and B, with or without indices, as metamathematical variables for formulas; the letter E, with or without indices, as

(¹) The operation of substitution will be contemplated only in the cases when the words $u_1, \cdots, u_k$ are mutually distinct and the occurrences of these words in F are mutually nonoverlapping.

metamathematical variables for elementary formulas; and the letters Γ, Δ, Θ and Λ, with or without indices, as metamathematical variables for lists of formulas.

The numbering of lemmas and theorems is continuous; the numbering of formulas is by section.

The basic results of this paper were announced at the Leningrad seminar on mathematical logic [7] and at the International Congress of Mathematicians of 1966 in Moscow [11]. Some of the results were announced at the jubilee scientific session of the Institute of Applied Mathematics of Tbilisi State University, which was held on the occasion of the fiftieth anniversary of the university in 1968 [12].

The author expresses deep thanks to G. E. Minc for formulation of the problem and valuable advice, as well as to N. A. Šanin for attention to the paper.

§1. Auxiliary calculi K_0 and $K_{0c}^{=}$

Let us denote by K_0 the modification in conformity with the language under consideration of the calculus C^* described in [13]. K_0 is a Gentzen variant of the classical predicate calculus, free of structural rules, in which all structural rules are admissible.(2) In the quantifier rules of this calculus there are no restrictions on the kind of terms substituted. Let us denote by K_{0c} the calculus obtained from K_0 by adding all structural rules (including the cut).

Let us denote by K_{0c}^* the calculus obtained from K_{0c} by adding the axiom $\rightarrow a_1 = a_1$ and the schema of axioms $\rightarrow a_1 = a_2 \supset ([A|_{a_1}^{\alpha}] \supset [A|_{a_2}^{\alpha}])$.

It is easy to see that K_{0c}^* and the classical predicate calculus with equality and function symbols are equiextensional (cf. [14], §73).

Let us denote by $K_{0c}^{=}$ the calculus obtained from K_{0c} by adding the schema of axioms $\Gamma \rightarrow \Delta_1, t = t, \Delta_2$ and the following rules for equality:

$$\frac{[\Gamma|_r^\alpha],\ r = s,\ [\Delta|_r^\alpha] \rightarrow [\Theta|_r^\alpha]}{[\Gamma|_s^\alpha],\ r = s,\ [\Delta|_s^\alpha] \rightarrow [\Theta|_s^\alpha]}, \quad \frac{[\Gamma|_r^\alpha],\ s = r,\ [\Delta|_r^\alpha] \rightarrow [\Theta|_r^\alpha]}{[\Gamma|_s^\alpha],\ s = r,\ [\Delta|_s^\alpha] \rightarrow [\Theta|_s^\alpha]}, \tag{1}$$

where α is an arbitrary free variable occurring in only one of the members of the list of formulas Γ, Δ, Θ.

In each of the rules (1), let us call the explicitly shown occurrences of the formulas $r = s$ and $s = r$ *auxiliary occurrences* of the rule under consideration, and let us call the occurrences of formulas in which replacement occurs *principal occurrences*. Let us call the formula $r = s$ or $s = r$ the *auxiliary formula* of the rule, and the formulas in which replacement occurs the *principal formulas*.

It is easy to prove the equiextensionality of K_{0c}^* and $K_{0c}^{=}$. It is sufficient for this to prove the following two assertions:

---

(2) A rule is called *admissible* in a calculus if the deducibility of its premises in that calculus implies the deducibility of its conclusion.

1) The sequents $\to a_1 = a_1$ and $\to a_1 = a_2 \supset ([A|^{\alpha}_{a_1}] \supset [A|^{\alpha}_{a_2}])$ are deducible in $K^{=}_{0c}$.

2) The sequent $\Gamma \to \Delta_1, t = t, \Delta_2$ is deducible in K^{*}_{0c}, and the rules (1) are derivable rules([3]) of K^{*}_{0c}.

Let us consider the axioms $\Gamma_1, A, \Gamma_2 \to \Delta_1, A, \Delta_2$ and $\Gamma \to \Delta_1, t = t, \Delta_2$ of $K^{=}_{0c}$. Let us say that the sequent $\Gamma_1, A, \Gamma_2 \to \Delta_1, A, \Delta_2$ (respectively, $\Gamma \to \Delta_1, t = t, \Delta_2$) is an axiom over elementary formulas if it is a prime sequent. It is easy to see that in an axiom over elementary formulas, the formula A is always an elementary formula.

Extending deduction upwards, as in the proof of Theorem 50 in [14], and using the direct invertibility (cf. [5]) of all rules of K_0, except $\forall\to$ and $\to\exists$, it is easy to prove the following lemma.

LEMMA 1. *Any deduction in $K^{=}_{0c}$ can be reconstructed into a deduction in $K^{=}_{0c}$ which has the same last sequent as the original deduction and in which all axioms are axioms over elementary formulas.*

Let us denote by $K^{=}_{0}$ the calculus obtained from $K^{=}_{0c}$ by removing all structural rules.

§2. Specialization of rules for equality

Let us call an application of a deductive rule *antecedent* (*succedent*) if the succedent (the antecedent) of the conclusion of the rule coincides with the succedent (with the antecedent) of each premise, i.e. if when applying this rule only the antecedents (the succedent) of sequents undergo change.

Let us call a deduction *antecedent* (*succedent*) if all applications of rules in it are antecedent (succedent).

Let us denote by J the auxiliary calculus whose postulates are, in the first place, the axiom schemata $\Gamma \to \Delta_1, t = t, \Delta_2$ and $\Gamma_1, A, \Gamma_2 \to \Delta_1, A, \Delta_2$ and, in the second place, all rules for equality.

In the sequel the symbol $E\to$ (respectively, $\to E$) will denote any antecedent (succedent) application of a rule for equality.

LEMMA 2. *Let a deduction of some sequent S in the calculus J be given in which the terminal application is an antecedent application of a rule and all remaining applications of rules are succedent. Then it is possible to construct a succedent deduction of this sequent in J.*

PROOF. The lemma is proved by induction on the height h of the given deduction.

([3]) A rule is called *derivable* in a calculus if its conclusion is deducible in that calculus from its premises.

If the given sequent S is an axiom, we replace the given deduction by this axiom. Let us assume in the sequel that S is not an axiom.

Basis: $h=1$ is trivial.

Induction step: $h>1$. Let us consider all possible cases.

Case 1. $h=2$. The terminal application of a rule is $E\rightarrow$. It may be assumed without loss of generality that an auxiliary formula of the rule is to the left of the principal one.

Let us write the deduction in the form

$$\frac{\Gamma_1,\ r=s,\ \Gamma_2,\ [A\,|^{\alpha}_{r}],\ \Gamma_3\rightarrow\Delta_1,\ [A\,|^{\alpha}_{r}],\ \Delta_2}{\Gamma_1,\ r=s,\ \Gamma_2,\ [A\,|^{\alpha}_{s}],\ \Gamma_3\rightarrow\Delta_1,\ [A\,|^{\alpha}_{r}],\ \Delta_2}\,E\rightarrow.$$

The following deduction possesses the required properties:

$$\frac{\Gamma_1,\ r=s,\ \Gamma_2,\ [A\,|^{\alpha}_{s}],\ \Gamma_3\rightarrow\Delta_1,\ [A\,|^{\alpha}_{s}],\ \Delta_2}{\Gamma_1,\ r=s,\ \Gamma_2,\ [A\,|^{\alpha}_{s}],\ \Gamma_3\rightarrow\Delta_1,\ [A\,|^{\alpha}_{r}],\ \Delta_2}\rightarrow E.$$

Case 2. $h>2$. Let us write the terminus of the given deduction in the form

$$\begin{array}{cl}\Gamma'\rightarrow\Delta' & \\ \hline \Gamma'\rightarrow\Delta & \rightarrow E \\ \hline \Gamma\rightarrow\Delta & E\rightarrow.\end{array}$$

Subcase 2.1. The principal occurrence of the application $E\rightarrow$ is not an auxiliary occurrence of $\rightarrow E$.

Let us replace the terminus of the given deduction by the configuration

$$\begin{array}{cl}\Gamma'\rightarrow\Delta' & \\ \hline \Gamma\rightarrow\Delta' & E\rightarrow \\ \hline \Gamma\rightarrow\Delta & \rightarrow E.\end{array}$$

By the induction hypothesis, a deduction of the sequent $\Gamma\rightarrow\Delta'$ can be reconstructed into the deduction D possessing the required properties.

Subcase 2.2. Case 2.1 does not hold. Without loss of generality the terminus of the given deduction can be written in the form

$$\begin{array}{cl}\Gamma_1,\ t=w,\ \Gamma_2,\ r=s,\ \Gamma_3\rightarrow\Delta_1,\ [A\,|^{\alpha}_{r}],\ \Delta_2 & \\ \hline \Gamma_1,\ t=w,\ \Gamma_2,\ r=s,\ \Gamma_3\rightarrow\Delta_1,\ [A\,|^{\alpha}_{s}],\ \Delta_2 & \rightarrow E \\ \hline \Gamma_1,\ t=w,\ \Gamma_2,\ \tilde{r}=\tilde{s},\ \Gamma_3\rightarrow\Delta_1,\ [A\,|^{\alpha}_{s}],\ \Delta_2 & E\rightarrow,\end{array}$$

where $t=w$ is an auxiliary formula of the application $E\rightarrow$ under consideration and the formula $\tilde{r}=\tilde{s}$ is obtained from $r=s$ as a result of this application.

Let us replace this part of the deduction by the configuration

$$\begin{array}{cl}\Gamma_1,\ t=w,\ \Gamma_2,\ r=s,\ \Gamma_3\rightarrow\Delta_1,\ [A\,|^{\alpha}_{r}],\ \Delta_2 & \\ \hline \Gamma_1,\ t=w,\ \Gamma_2,\ \tilde{r}=\tilde{s},\ \Gamma_3\rightarrow\Delta_1,\ [A\,|^{\alpha}_{r}],\ \Delta_2 & E\rightarrow \\ \hline \Gamma_1,\ t=w,\ \Gamma_2,\ \tilde{r}=\tilde{s},\ \Gamma_3\rightarrow\Delta_1,\ [A\,|^{\alpha}_{\tilde{r}}],\ \Delta_2 & \rightarrow E \\ \hline \Gamma_1,\ t=w,\ \Gamma_2,\ \tilde{r}=\tilde{s},\ \Gamma_3\rightarrow\Delta_1,\ [A\,|^{\alpha}_{\tilde{s}}],\ \Delta_2 & \rightarrow E \\ \hline \Gamma_1,\ t=w,\ \Gamma_2,\ \tilde{r}=\tilde{s},\ \Gamma_3\rightarrow\Delta_1,\ [A\,|^{\alpha}_{s}],\ \Delta_2 & \rightarrow E.\end{array}$$

Here one of the applications $\to E$ can be an identity step (then we eliminate the superfluous sequents from this configuration).

By the induction hypothesis, a deduction of the sequent $\Gamma_1, t = w, \Gamma_2, \widetilde{r} = \widetilde{s}, \Gamma_3 \to \Delta_1, [A|_r^\alpha], \Delta_2$ can be reconstructed into the deduction D possessing the required properties. Realizing this reconstruction, we obtain the deduction of the sequent S in J.

LEMMA 3. *With respect to any deduction of any sequent in the calculus J, it is possible to construct a succedent deduction of this sequent in J.*

PROOF. With the use of Lemma 2, this lemma is proved by induction on the number n of antecedent applications of rules in a given deduction.

Let us call a deduction D of the sequent $\Gamma \to \Delta$ in the calculus J *weeded* (cf. [15]) if the principal formula of each application of a rule for equality in D is traceable to an axiom [13].

LEMMA 4. *Any deduction D of the sequent $\Gamma \to \Delta$ in J can be reconstructed into a weeded succedent deduction D' of the same sequent in J.*

PROOF. First let us construct with the help of Lemma 3 a succedent deduction of the sequent $\Gamma \to \Delta$ in J. Then after sorting out all occurrences of the formulas of the list Δ, we find the occurrence of formula A of the list Δ which is traceable to an axiom. Let us write Δ as Δ_1, A, Δ_2, and let us write the ith sequent from the top of the new deduction in the form $\Gamma \to \Delta_1^i, A^i, \Delta_2^i$, where $\Delta_1^i, A^i, \Delta_2^i$ are ancestors of Δ_1, A, Δ_2, respectively [16], $i = 1, \cdots, h$ (h is the height of the new deduction).

It is obvious in this connection that $\Delta_1^h \mathrel{\overset{\circ}{=}} \Delta_1$, $\Delta_2^h \mathrel{\overset{\circ}{=}} \Delta_2$ and $A^h \mathrel{\overset{\circ}{=}} A$. Let us replace the ith sequent from the top of the given deduction by the sequent $\Gamma \to \Delta_1, A^i, \Delta_2$ and cancel repetitions of sequents. The deduction obtained will be the one desired. The lemma is proved.

Let us call an application of a rule for equality *left-sided* (*right-sided*) if the principal formula of that application is an equality and when applying that rule only the left (the right) side of this equality undergoes change.

Let us call an application of a rule for equality *two-sided* if the principal formula of that application is an equality and when applying that rule both the left and the right sides of this equality undergo change. It is easy to see that every two-sided application of a rule for equality can be replaced by two applications of rules for equality, a left-sided and a right-sided.

Let us call a deduction in J *left-sided* (*right-sided*) if no application of rules for equality in this deduction is right-sided (respectively, left-sided).

It is easy to see that any axiom of the form $\Gamma_1, r = s, \Gamma_2 \to \Delta_1, r = s, \Delta_2$ can be replaced in a deduction in $K_{0c}^{=}$ by, for example, the configuration

$$\frac{\Gamma_1,\ r = s,\ \Gamma_2 \to \Delta_1,\ r = r,\ \Delta_2}{\Gamma_1,\ r = s,\ \Gamma_2 \to \Delta_1,\ r = s,\ \Delta_2} \to E.$$

LEMMA 5. *Let a weeded deduction of the sequent* $\Gamma \to \Delta_1,\ u = v,\ \Delta_2$ *in the calculus* J *be given in which the explicity shown occurrence of the formula* $u = v$ *in the sequent* $\Gamma \to \Delta_1,\ u = v,\ \Delta_2$ *is traceable to an axiom. Then it is possible to construct a weeded succedent right-sided (left-sided) deduction of that sequent in* J *in which* $\Gamma \to \Delta_1,\ u = u,\ \Delta_2$ *(respectively,* $\Gamma \to \Delta_1,\ v = v,\ \Delta_2$*) is the highest sequent.*

The proof of this assertion is analogous to the proof of Lemma 3.

LEMMA 6. *Let a weeded-deduction of the sequent* $\Gamma_1, P(t_1, \cdots, t_n), \Gamma_2 \to \Delta_1, P(w_1, \cdots, w_n), \Delta_2$ *in the calculus* J *be given in which* $\Gamma_1, P(t_1, \cdots, t_n), \Gamma_2 \to \Delta_1, P(t_1, \cdots, t_n), \Delta_2$ *is the highest sequent. Then it is possible to construct a weeded antecedent deduction of the sequent* $\Gamma_1, P(t_1, \cdots, t_n), \Gamma_2 \to \Delta_1, P(w_1, \cdots, w_n), \Delta_2$ *in* J *in which* $\Gamma_1, P(w_1, \cdots, w_n), \Gamma_2 \to \Delta_1, P(w_1, \cdots, w_n), \Delta_2$ *is the highest sequent.*

The proof of this assertion is analogous to the proof of Lemma 3.

An antecedent deduction cannot be constructed in J with respect to every deduction of the same sequent in J. For example, a deduction of the sequent $a_1 = a_2 \to f(a_1) = f(a_2)$ in J cannot be reconstructed into a deduction possessing the required properties.

§3. Elimination of structural rules from $K_{0c}^{=}$

Lemma 3, which makes it possible to restrict ourselves to succedent applications of rules for equality, admits numerous applications. As an illustration of the possibility of applying Lemma 3, let us prove a theorem concerning the elimination of the cut from $K_{0c}^{=}$. A similar assertion was cited for the first time (without proof) in [6]. A proof of a theorem concerning the elimination of the cut from a calculus similar to $K_{0c}^{=}$ is cited in [17].

The lemmas proved below bear a preparatory character. They will be utilized for the elimination of structural rules from $K_{0c}^{=}$.

LEMMA 7. *A deduction of the sequent* $\Gamma_1,\ t = t,\ \Gamma_2 \to \Delta$ *in the calculus* J *can be constructed with respect to any deduction of the sequent* $\Gamma_1, \Gamma_2 \to \Delta$ *in* J.

PROOF. Let us, with the help of Lemma 3, construct in J a succedent deduction of the sequent $\Gamma_1,\ t = t,\ \Gamma_2 \to \Delta$, and then let us cancel in each of the antecedents of the sequents encountered in the new deduction all occurrences of formulas which are ancestors of the explicitly shown occurrence of the formula $t = t$ in the given sequent $\Gamma_1, t = t,\ \Gamma_2 \to \Delta$ as well as repetitions of sequents. The configuration obtained will be the desired deduction.

Let us describe a process which permits obtaining from some deduction in J two configurations which are deductions in J from certain hypotheses.

Let a weeded succedent right-sided deduction D of the sequent $\Gamma_1, \Gamma_2 \to \Delta_1$,

$r = s,\ \Delta_2$ in J be given in which the explicitly shown occurrence of the formula $r = s$ is traceable to an axiom.

Let us write D in the form

$$\frac{\Gamma_1,\ \Gamma_2 \to \Delta_1,\ r = r,\ \Delta_2}{\Gamma_1,\ \Gamma_2 \to \Delta_1,\ r = s_1,\ \Delta_2} \to E$$

$$\cdots$$

$$\frac{\Gamma_1,\ \Gamma_2 \to \Delta_1,\ r = s_{l-1},\ \Delta_2}{\Gamma_1,\ \Gamma_2 \to \Delta_1,\ r = s_l,\ \Delta_2} \to E$$

$$\cdots$$

$$\frac{\Gamma_1,\ \Gamma_2 \to \Delta_1,\ r = s_n,\ \Delta_2}{\Gamma_1,\ \Gamma_2 \to \Delta_1,\ r = s,\ \Delta_2} \to E,$$

where the term s_1 is obtained from the term $r, \cdots$, the term s_l is obtained from $s_{l-1}, \cdots$, and the term s is obtained from the term s_n as the result of applying the rule $\to E$.

Let us denote by $D^*_{A,\alpha}$ the configuration

$$\frac{\Gamma_1,\ [A\,|^{\alpha}_{r}],\ \Gamma_2 \to \Delta_1,\ \Delta_2}{\Gamma_1,\ [A\,|^{\alpha}_{s_1}],\ \Gamma_2 \to \Delta_1,\ \Delta_2} E \to$$

$$\cdots$$

$$\frac{\Gamma_1,\ [A\,|^{\alpha}_{s_{l-1}}],\ \Gamma_2 \to \Delta_1,\ \Delta_2}{\Gamma_1,\ [A^{\alpha}_{s_l}],\ \Gamma_2 \to \Delta_1,\ \Delta_2} E \to$$

$$\cdots$$

$$\frac{\Gamma_1,\ [A\,|^{\alpha}_{s_n}],\ \Gamma_2 \to \Delta_1,\ \Delta_2}{\Gamma_1,\ [A\,|^{\alpha}_{s}],\ \Gamma_2 \to \Delta_1,\ \Delta_2} E \to,$$

where A is an arbitrary formula and α is an arbitrary free variable.

$D^*_{A,\alpha}$ is a deduction of the sequent $\Gamma_1, [A|^{\alpha}_{s}], \Gamma_2 \to \Delta_1, \Delta_2$ from the sequent $\Gamma_1, [A|^{\alpha}_{r}], \Gamma_2 \to \Delta_1, \Delta_2$ in J.

Analogously defined is the configuration $D^*_{A,\alpha}$, which is a deduction of the sequent $\Gamma_1, [A|^{\alpha}_{r}], \Gamma_2 \to \Delta_1, \Delta_2$ from the sequent $\Gamma_1, [A|^{\alpha}_{s}], \Gamma_2 \to \Delta_1, \Delta_2$ in J.

It is easy to see that the heights of the configurations $D, D^*_{A,\alpha}$ and $D^{**}_{A,\alpha}$ are equal.

LEMMA 8. *If the sequents* $\Gamma_1, \Gamma_2 \to \Delta_1, r = s, \Delta_2$ *and* $\Gamma_1, f(t_1, \cdots, t_{i-1}, r, t_{i+1}, \cdots, t_n) = f(t_1, \cdots, t_{i-1}, s, t_{i+1}, \cdots, t_n), \Gamma_2 \to \Delta_1, \Delta_2$ *are deducible in the calculus* J, *so is the sequent* $\Gamma_1, \Gamma_2 \to \Delta_1, \Delta_2$.

PROOF. With the help of Lemma 4 let us construct a succedent weeded deduction D_0 of the sequent $\Gamma_1, \Gamma_2 \to \Delta_1,\ r = s,\ \Delta_2$ in J. If the explicitly shown occurrence of the formula $r = s$ in this sequent is not traceable to an axiom, let us cancel in each of the succedents of the sequents encountered in D_0 all occurrences of formulas which are ancestors of the explicitly shown occurrence of the formula $r = s$. The deduction obtained will be a deduction of the sequent $\Gamma_1, \Gamma_2 = \Delta_1, \Delta_2$ in J. If in D_0 the explicitly shown occurrence of the formula $r = s$ is traceable to an axiom, let us, with the help of Lemma 5, reconstruct D_0 into a weeded-succedent left-sided deduction D in J in which $\Gamma_1, \Gamma_2 \to \Delta_1,\ s=s,\ \Delta_2$ is the highest sequent. Let us denote by D_1 a deduction of the sequent $\Gamma_1, f(t_1, \cdots, t_{i-1}, r, t_{i+1}, \cdots, t_n) = f(t_1, \cdots, t_{i-1}, s, t_{i+1}, \cdots, t_n),\ \Gamma_2 \to \Delta_1, \Delta_2$ in J. The following configuration will be a deduction of the sequent $\Gamma_1, f(t_1, \cdots, t_{i-1}, r, t_{i+1}, \cdots, t_n) = f(t_1, \cdots, t_{i-1}, r, t_{i+1}, \cdots, t_n),\ \Gamma_2 \to \Delta_1, \Delta_2$ in J:

$$\begin{array}{l} D_1 \left\{ \dfrac{\cdots}{\Gamma_1,\ f(t_1,\ \ldots,\ t_{i-1},\ r,\ t_{i+1},\ \ldots,\ t_n) = f(t_1,\ \ldots,\ t_{i-1},\ s,\ t_{i+1},\ \ldots,\ t_n),\ \Gamma_2 \to \Delta_1,\ \Delta_2} \right. \\ D^{**}_{A,\, a_1} \left\{ \dfrac{\cdots}{\Gamma_1,\ f(t_1,\ \ldots,\ t_{i-1},\ r,\ t_{i+1},\ \ldots,\ t_n) = f(t_1,\ \ldots,\ t_{i-1},\ r,\ t_{i+1},\ \ldots,\ t_n),\ \Gamma_2 \to \Delta_1,\ \Delta_2} \right. , \end{array}$$

where A denotes the formula

$$f(t_1,\ \ldots,\ t_{i-1},\ a_1,\ t_{i+1},\ \ldots,\ t_n) = f(t_1,\ \ldots,\ t_{i-1},\ a_1,\ t_{i+1},\ \ldots,\ t_n).$$

Applying Lemma 7 to the lower sequent of this deduction, we construct a deduction of the sequent $\Gamma_1, \Gamma_2 \to \Delta_1, \Delta_2$ in J.

LEMMA 9. *If the sequents* $\Gamma_1, \Gamma_2 \to \Delta_1, r=s,\ \Delta_2$; $\Gamma_1, \Gamma_2 \to \Delta_1,\ r=t,\ \Delta_2$ *and* $\Gamma_1, s=t,\ \Gamma_2 \to \Delta_1, \Delta_2$ *are deducible in* J, *so is the sequent* $\Gamma_1, \Gamma_2 \to \Delta_1, \Delta_2$.

PROOF. First with the help of Lemma 4 let us reconstruct, respectively, the given deductions of the sequents $\Gamma_1, \Gamma_2 \to \Delta_1,\ r=s,\ \Delta_2$ and $\Gamma_1, \Gamma_2 \to \Delta_1,\ r=t,\ \Delta_2$ into the weeded deductions D_1 and D_2 in J. If the explicitly shown occurrence of the formula $r=s$ in the sequent $\Gamma_1, \Gamma_2 \to \Delta_1, r=s,\ \Delta_2$ (respectively, $r=t$ in the sequent $\Gamma_1, \Gamma_2 \to \Delta_1,\ r=t,\ \Delta_2)$ is not traceable in D_1 (in D_2) to an axiom, let us cancel in each of the succedents of the sequents encountered in D_1 (in D_2) all occurrences of formulas which are ancestors of the explicitly shown occurrence of the formula $r=s$ (of $r=t$). The configuration obtained will be the desired deduction of the sequent $\Gamma_1, \Gamma_2 \to \Delta_1, \Delta_2$ in J.

If the occurrences mentioned are traceable to axioms, let us, with the help of Lemma 5, reconstruct the deduction D_1 (the deduction D_2) into a weeded succedent right-sided deduction D_3 (weeded succedent left-sided deduction D_4) in J in which $\Gamma_1, \Gamma_2 \to \Delta_1,\ r=r,\ \Delta_2$ (respectively, $\Gamma_1, \Gamma_2 \to \Delta_1,\ t=t,\ \Delta_2$) is the highest sequent.

Let us denote by D_5 a deduction of the sequent $\Gamma_1,\ s=t,\ \Gamma_2 \to \Delta_1, \Delta_2$ in J. The following configuration is a deduction of $\Gamma_1,\ s=s,\ \Gamma_2 \to \Delta_1, \Delta_2$ in J:

$$\begin{array}{l} D_5 \left\{ \dfrac{\cdots}{\Gamma_1,\ s=t,\ \Gamma_2 \to \Delta_1,\ \Delta_2} \right. \\ (D_4)^{**}_{s=a_1,\, a_1} \left\{ \dfrac{\cdots}{\Gamma_1,\ s=r,\ \Gamma_2 \to \Delta_1,\ \Delta_2} \right. \\ (D_3)^{*}_{a_1=a_1,\, a_1} \left\{ \dfrac{\cdots}{\Gamma_1,\ s=s,\ \Gamma_2 \to \Delta_1,\ \Delta_2} \right. . \end{array}$$

Applying Lemma 7 to the last sequent of this deduction, we construct a deduction of the sequent $\Gamma_1, \Gamma_2 \rightarrow \Delta_1, \Delta_2$ in J.

LEMMA 10. *If the sequents*

$$\Gamma_1, \Gamma_2 \rightarrow \Delta_1, r = s, \Delta_2; \Gamma_1, \Gamma_2 \rightarrow \Delta_1, P(t_1, \ldots, t_{i-1}, r, t_{i+1}, \ldots, t_n), \Delta_2$$

and

$$\Gamma_1, P(t_1, \ldots, t_{i-1}, s, t_{i+1}, \ldots, t_n), \Gamma_2 \rightarrow \Delta_1, \Delta_2,$$

are deducible in the calculus J, so is $\Gamma_1, \Gamma_2 \rightarrow \Delta_1, \Delta_2$.

PROOF. First with the help of Lemma 4 let us reconstruct the given deductions of the sequents $\Gamma_1, \Gamma_2 \rightarrow \Delta_1, r = s, \Delta_2$ and $\Gamma_1, \Gamma_2 \rightarrow \Delta_1, P(t_1, \cdots, t_{i-1}, r, t_{i+1}, \cdots, t_n), \Delta_2$, into the weeded deductions D_1 and D_2 in J. Let us assume that the explicitly shown occurrence of the formula $r = s$ in the sequent $\Gamma_1, \Gamma_2 \rightarrow \Delta_1, r = s, \Delta_2$ (respectively, of $P(t_1, \cdots, t_{i-1}, r, t_{i+1}, \cdots, t_n)$ in the sequent $\Gamma_1, \Gamma_2 \rightarrow \Delta_1, P(t_1, \cdots, t_{i-1}, r, t_{i+1}, \cdots, t_n)\ \Delta_2$) is traceable in D_1 (in D_2) to an axiom; otherwise the desired deduction is easily constructed. With the help of Lemma 5, let us reconstruct D_1 into a weeded succedent right-sided deduction D_3 in J in which $\Gamma_1, \Gamma_2 \rightarrow \Delta_1, r = r, \Delta_2$ is the highest sequent.

Since the explicitly shown occurrence of the formula $P(t_1, \cdots, t_{i-1}, r, t_{i+1}, \cdots, t_n)$ is traceable in D_2 to an axiom, there is an elementary formula $P(w_1, \cdots, w_n)$ occurring in Γ_1, Γ_2 such that the highest sequent of the deduction D_2 will be Γ_3, $P(w_1, \cdots, w_n), \Gamma_4 \rightarrow \Delta_1, P(w_1, \cdots, w_n), \Delta_2$, where the list Γ_1, Γ_2 coincides with the list $\Gamma_3, P(w_1, \cdots, w_n), \Gamma_4$, with the help of Lemma 6 we reconstruct D_2 into an antecedent deduction D_4 in J in which $\Gamma_3, P(t_1, \cdots, t_{i-1}, r, t_{i+1}, \cdots, t_n)$, $\Gamma_4 \rightarrow \Delta_1, P(t_1, \cdots, t_{i-1}, r, t_{i+1}, \cdots, t_n), \Delta_2$ is the highest sequent.

With the help of Lemmas 3 and 4, let us construct a weeded succedent deduction D_5 of the sequent $\Gamma_1, P(t_1, \cdots, t_{i-1}, s, t_{i+1}, \cdots, t_n), \Gamma_2 \rightarrow \Delta_1, \Delta_2$ in J. Let us consider two cases.

Case 1. The formula $P(t_1, \cdots, t_{i-1}, s, t_{i+1}, \cdots, t_n)$ does not occur in the succedent of the highest sequent of D_5. Then in each of the antecedents of the sequents encountered in D_5 we cancel all occurrences of formulas which are ancestors of the explicitly shown occurrence of $P(t_1, \cdots, t_{i-1}, s, t_{i+1}, \cdots, t_n)$ in the sequent $P(t_1, \cdots, t_{i-1}, s, t_{i+1}, \cdots, t_n)$. The configuration obtained will be a deduction of $\Gamma_1, \Gamma_2 \rightarrow \Delta_1, \Delta_2$ in J.

Case 2. There is an elementary formula $P(u_1, \cdots, u_n)$ occurring in Δ_1, Δ_2 such that $\Gamma_1, P(t_1, \cdots, t_{i-1}, s, t_{i+1}, \cdots, t_n), \Gamma_2 \rightarrow \Delta_3, P(t_1, \cdots, t_{i-1}, s, t_{i+1}, \cdots, t_n)$, Δ_4 is the highest sequent of deduction D_5 and the explicitly shown succedent occurrence of the formula $P(t_1, \cdots, t_{i-1}, s, t_{i+1}, \cdots, t_n)$ is an ancestor of the occurrence of the formula $P(u_1, \cdots, u_n)$ in the list Δ_1, Δ_2. Then $\Gamma_1, \Gamma_2 \rightarrow \Delta_1, \Delta_2$ can be written in the form

$$\Gamma_3,\ P(w_1,\ \ldots,\ w_n),\ \Gamma_4 \rightarrow \Delta_3,\ P(u_1,\ \ldots,\ u_n),\ \Delta_4.$$

Then it is possible to construct the following deduction of the sequent $\Gamma_1, \Gamma_2 \rightarrow \Delta_1, \Delta_2$ in J:

$$D'_3 \left\{ \begin{array}{c} \underline{\Gamma_3,\ P(t_1,\ \ldots,\ t_{i-1},\ r,\ t_{i+1},\ \ldots,\ t_n),\ \Gamma_4 \rightarrow \Delta_3,\ P(t_1,\ \ldots,\ t_{i-1},\ r,\ t_{i+1},\ \ldots,\ t_n),\ \Delta_4} \\ \ldots \end{array} \right.$$

$$D'_4 \left\{ \begin{array}{c} \overline{\Gamma_3,\ P(t_1,\ \ldots,\ t_{i-1},\ r,\ t_{i+1},\ \ldots,\ t_n),\ \Gamma_4 \rightarrow \Delta_3,\ P(t_1,\ \ldots,\ t_{i-1},\ s,\ t_{i+1},\ \ldots,\ t_n),\ \Delta_4} \\ \ldots \\ \hline \Gamma_3,\ P(w_1,\ \ldots,\ w_n),\ \Gamma_4 \rightarrow \Delta_3,\ P(t_1,\ \ldots,\ t_{i-1},\ s,\ t_{i+1},\ \ldots,\ t_n),\ \Delta_4 \end{array} \right.$$

$$D'_5 \left\{ \begin{array}{c} \ldots \\ \hline \Gamma_3,\ P(w_1,\ \ldots,\ w_n),\ \Gamma_4 \rightarrow \Delta_3,\ P(u_1,\ \ldots,\ u_n),\ \Delta_4 \end{array} \right. ,$$

where the configurations D'_3, D'_4 and D'_5 are obtained in an obvious way from D_3, D_4 and D_5, respectively.

Let us call (following [18]) an application of a rule for equality in some deduction *regular* if only rules for equality are applied above it. Otherwise we call the application *irregular*. We call a deduction *regular* if all applications in it of rules for equality are regular.

Let us denote by J^* the calculus obtained from J by adding the rule $\supset\rightarrow$.

The following is proved analogously to the proof of Lemma 2.2.2 of [18].

LEMMA 11. *With respect to any deduction of some sequent S in the calculus J^*, a regular deduction can be constructed of that sequent in J^*.*

If $P_1, \cdots, P_l$ are predicate symbols and $f_1, \cdots, f_k$ are function symbols, the expression $\mathrm{Eq}\,(=, P_1, \cdots, P_l, f_1, \cdots, f_k)$ has the same meaning as in §73 of [14], i.e. it denotes the conjunction of closed formulas expressing the reflexivity and transitivity of equality, as well as the possibility of replacing equal objects by each other in any argument places of $P_1, \cdots, P_l, f_1, \cdots, f_k$. Let us denote by $\overline{\mathrm{Eq}}\,(=, P_1, \cdots, P_l, f_1, \cdots, f_k)$ the list of closed formulas obtained from $\mathrm{Eq}\,(=, P_1, \cdots, P_l, f_1, \cdots, f_k)$ as the result of replacing all occurrences of conjunction by commas.

Let us consider the following types of formulas:

$$t = t, \tag{b_0}$$

$$t_1 = t_2 \supset (t_1 = t_3 \supset t_2 = t_3), \tag{b_1}$$

$$(v = w) \supset (P(w_1,\ \ldots,\ w_{i-1},\ v,\ w_{i+1},\ \ldots,\ w_n) \supset P(w_1,\ \ldots,\ w_{i-1},\ w,\ w_{i+1},\ \ldots,\ w_n)), \tag{b_2}$$

$$(v = w) \supset (f(u_1,\ \ldots,\ u_{i-1},\ v,\ u_{i+1},\ \ldots,\ u_m) = f(u_1,\ \ldots,\ u_{i-1},\ w,\ u_{i+1},\ \ldots,\ u_m)). \tag{b_3}$$

Formulas of types (b_0) and (b_1) (formulas (b_2) and (b_3)) are obtained as the result of substituting terms into open axioms of equality for = (respectively for P and f) (cf. [14]) in place of their free variables.

LEMMA 12. *Let $\Gamma_1, \Gamma_2 \to \Delta$ be an arbitrary prime sequent. Then with respect to any deduction of the sequent*

$$\Gamma_1,\ t_1 = t_2 \supset (t_1 = t_3 \supset t_2 = t_3),\ \Gamma_2 \to \Delta \tag{1}$$

in J^*, it is possible to construct a deduction of the sequent $\Gamma_1, \Gamma_2 \to \Delta$ in J.

PROOF. First with the help of Lemma 1 we construct a deduction D of the sequent (1) in J^* in which all axioms are prime. Then with the help of Lemma 11 we reconstruct D into a regular deduction D' in J^*. We assume that the explicitly shown occurrence of a formula of the form (b_1) is traceable to an axiom. Since the sequent $\Gamma_1, \Gamma_2 \to \Delta$ is prime, the terminus of the deduction D' has the form

$$\dfrac{\Gamma_1,\ \Gamma_2 \to \Delta,\ t_1 = t_2;\ \ \Gamma_1,\ t_1 = t_3 \supset t_2 = t_3,\ \Gamma_2 \to \Delta \quad \dfrac{\Gamma_1,\ \Gamma_2 \to \Delta,\ t_1 = t_3;\ \ \Gamma_1,\ t_2 = t_3,\ \Gamma_2 \to \Delta}{} \supset\to}{\Gamma_1,\ t_1 = t_2 \supset (t_1 = t_3 \supset t_2 = t_3),\ \Gamma_2 \to \Delta} \supset\to .$$

Applying Lemma 9 to the sequents $\Gamma_1, \Gamma_2 \to \Delta,\ t_1 = t_2$; $\Gamma_1, \Gamma_2 \to \Delta,\ t_1 = t_3$ and $\Gamma_1,\ t_2 = t_3,\ \Gamma_2 \to \Delta$, which are deducible in J, we construct a deduction of the sequent $\Gamma_1, \Gamma_2 \to \Delta$ in J.

LEMMA 13. *Let $\Gamma_1, \Gamma_2 \to \Delta$ be an arbitrary prime sequent. Then with respect to any deduction of the sequent*

$$\begin{gathered}\Gamma_1,\ (v = w) \supset (P(w_1, \ldots, w_{i-1}, v, w_{i+1}, \ldots, w_n) \\ \supset P(w_1, \ldots, w_{i-1}, w, w_{i+1}, \ldots, w_n)),\ \Gamma_2 \to \Delta\end{gathered} \tag{2}$$

in J^*, it is possible to construct a deduction of the sequent $\Gamma_1, \Gamma_2 \to \Delta$ in J.

PROOF. Analogously to the proof of Lemma 12, let us apply Lemma 10 to the upper sequents of the configuration

$$\dfrac{\begin{gathered}\Gamma_1,\ \Gamma_2 \to \Delta,\ v = w;\ \ \Gamma_1,\ P(w_1, \ldots, w_{i-1}, v, w_{i+1}, \ldots, w_n) \\ \supset P(w_1, \ldots, w_{i-1}, w, w_{i+1}, \ldots, w_n),\ \Gamma_2 \to \Delta\end{gathered} \quad \dfrac{\begin{gathered}\Gamma_1,\ \Gamma_2 \to \Delta,\ P(w_1, \ldots, w_{i-1}, v, w_{i+1}, \ldots, w_n); \\ \Gamma_1,\ P(w_1, \ldots, w_{i-1}, w, w_{i+1}, \ldots, w_n),\ \Gamma_2 \to \Delta\end{gathered}}{} \supset\to}{\begin{gathered}\Gamma_1,\ (v = w) \supset (P(w_1, \ldots, w_{i-1}, v, w_{i+1}, \ldots, w_n) \\ \supset P(w_1, \ldots, w_{i-1}, w, w_{i+1}, \ldots, w_n)),\ \Gamma_2 \to \Delta\end{gathered}} \supset\to .$$

LEMMA 14. *Let $\Gamma_1, \Gamma_2 \to \Delta$ be an arbitrary prime sequent. Then with respect to any deduction of the sequent*

$$\begin{gathered}\Gamma_1,\ (v = w) \supset (f(u_1, \ldots, u_{i-1}, v, u_{i+1}, \ldots, u_m) \\ = f(u_1, \ldots, u_{i-1}, w, u_{i+1}, \ldots, u_m)),\ \Gamma_2 \to \Delta\end{gathered} \tag{3}$$

in J^*, *it is possible to construct a deduction of the sequent* $\Gamma_1, \Gamma_2 \to \Delta$ *in* J.

The proof is analogous to that of Lemma 12, with the use of Lemma 8.

Let $A_1, \cdots, A_m$ be formulas belonging to one of the types $(b_0), (b_1), (b_2)$ or (b_3), and let $\Gamma_1, \Gamma_2 \to \Delta$ be a prime sequent. Let us divide the occurrences of logical symbols in $\Gamma_1, A_1, \cdots, A_m, \Gamma_2 \to \Delta$ into $m + 1$ classes $C_1, \cdots, C_{m+1}$. (We say that C_i is "higher" than C_j if $i < j$.) We attribute to the first class all occurrences of logical symbols in A_1, to the second class all occurrences of logical symbols in $A_2, \cdots$, to the mth class all occurrences of logical symbols in A_m, and to the $(m + 1)$th class, all occurrences of logical symbols in Γ_1, Γ_2 and Δ.

Let us call such a partition of occurrences of logical symbols the *canonical partition*(4) for the sequent $\Gamma_1, A_1, \cdots, A_m, \Gamma_2 \to \Delta$.

Let $\mathfrak{C}$ be the canonical partition for the sequent $\Gamma_1, A_1, \cdots, A_m, \Gamma_2 \to \Delta$, and let D be a deduction of that sequent in J^*. We say that the deduction D is consistent with the partition $\mathfrak{C}$ if in any branch of the deduction D applications of logical rules belonging to occurrences in a higher class are placed higher than occurrences of logical rules belonging to occurrences of a lower class.

LEMMA 15. *Let* $A_1, \cdots, A_m$ *be formulas of one of the types* $(b_0), (b_1), (b_2)$ *or* (b_3), *let* $\Gamma_1, \Gamma_2 \to \Delta$ *be a prime sequent, and let* $\mathfrak{C}$ *be the canonical partition for the sequent* $\Gamma_1, A_1, \cdots, A_m, \Gamma_2 \to \Delta$. *Let a regular deduction* D *of the sequent* $\Gamma_1, A_1, \cdots, A_m, \Gamma_2 \to \Delta$ *in* J^* *be given which is consistent with the partition* $\mathfrak{C}$ *and in which all axioms are axioms over elementary formulas. Then it is possible to construct a deduction of the sequent* $\Gamma_1, \Gamma_2 \to \Delta$ *in* J.

PROOF. The lemma is proved by induction on the number m. The basis $m = 0$ is trivial.

Induction step: $m > 0$. Let us write as $S_1, \cdots, S_r$ the complete list of occurrences of sequents which are the conclusions of the lowest applications of the rules of C_1 in D. For each $i = 1, \cdots, r$ the base of the occurrence S_i has the form $\Gamma^i, A_1, \Theta^i \to \Delta^i$, where $\Gamma^i, \Theta^i \to \Delta^i$ is a prime sequent. Let us denote by D_i a deduction terminating with the occurrence S_i. With the help of Lemmas 7, 12, 13 and 14 we construct the deduction D_i' of the sequent $\Gamma^i, \Theta^i \to \Delta^i$ $(i = 1, \cdots, r)$ in J. In D we replace each deduction D_i by D_i', and cancel all explicitly shown occurrences of the formula A_1 in the sequent $\Gamma_1, A_1, \cdots, A_m, \Gamma_2 \to \Delta$. The configuration obtained will be a regular deduction of $\Gamma_1, A_2, \cdots, A_m, \Gamma_2 \to \Delta$ in J^* which is consistent with the canonical partition of that sequent and in which all axioms are axioms over elementary formulas.

By the induction hypothesis it is possible to construct a deduction of $\Gamma_1, \Gamma_2 \to \Delta$ in J.

(4) Let us note that the canonical partition is a classical regular partition (cf. [13]).

Let $\mathfrak{C}$ be any classically regular partition (cf. [13]) for the sequent $\Gamma \to \Delta$. Let us say that the sequent $\Gamma \to \Delta$ is *normally deducible* in $K_0^=$ relative to $\mathfrak{C}$ if it is possible to construct a regular deduction of the sequent $\Gamma \to \Delta$ in $K_0^=$ which is consistent with $\mathfrak{C}$ and in which all axioms are axioms over elementary formulas.

LEMMA 16. *Let $P_1, \cdots, P_l, f_1, \cdots, f_k$ be the complete list of predicate and function symbols occurring in the sequent $\Gamma \to \Delta$. Let $\mathfrak{C}$ be any classically regular partition (cf. [13]) for the sequent $\Gamma \to \Delta$. The sequent $\Gamma \to \Delta$ is normally deducible in $K_0^=$ relative to $\mathfrak{C}$ if and only if the sequent*

$$\bar{\mathrm{E}}\mathrm{q}\,(=,\ P_1,\ \ldots,\ P_l,\ f_1,\ \ldots,\ f_k),\ \Gamma \to \Delta \tag{4}$$

is deducible in K_0.

PROOF. If the sequent $\Gamma \to \Delta$ is normally deducible in $K_0^=$ relative to $\mathfrak{C}$, sequent (4) is deducible in the classical predicate calculus K_0 (cf. Theorem 41 of [14]).

Now let a deduction of (4) in K_0 be given. Let $\mathfrak{C}$ be a classically regular partition for the sequent $\Gamma \to \Delta$, and let $C_3, \cdots, C_{m+2}$ be the enumeration of all classes of $\mathfrak{C}$ (the number of classes in $\mathfrak{C}$ is m). We prove that it is then possible to construct a regular deduction of the sequent $\Gamma \to \Delta$ in $K_0^=$ which is consistent with $\mathfrak{C}$.

With the help of Lemma 1 let us construct a deduction of (4) in K_0 in which all axioms are axioms over elementary formulas.

We divide the occurrences of logical symbols in (4) into the $m+2$ classes $C_1, \cdots, C_{m+2}$. We say that C_i is "higher" than C_j if $i<j$. Let us attribute to the first class all occurrences in $\bar{\mathrm{E}}\mathrm{q}\,(=, P_1, \cdots, P_l, f_1, \cdots, f_k)$ of the logical symbol $\supset$, and to the second class all occurrences in $\bar{\mathrm{E}}\mathrm{q}\,(=, P_1, \cdots, P_l, f_1, \cdots, f_k)$; $C_3, C_4, \cdots, C_{m+2}$ of the logical symbol $\forall$; $C_3, \cdots, C_{m+2}$ is the enumeration of all classes in the partition $\mathfrak{C}$ for the sequent $\Gamma \to \Delta$. The latter partition is classically regular. Consequently any deduction of sequent (4) will be classically almost consistent with the indicated partition.

We construct a deduction D_0 of (4) in K_0 which satisfies the condition of Theorem 3.1 of [13], and with the help of the same Theorem 3.1 we reconstruct D_0 into a deduction D in K_0 which is consistent with this partition.

Let us write in the form $S_1, \cdots, S_n$ the complete list of occurrences of sequents which are conclusions of the lowest applications of rules of C_1 in D. For each $i = 1, \cdots, n$ the base of the occurrence S_i has the form

$$\Gamma_i',\ A_1,\ \ldots,\ A_r,\ \Gamma_i'' \to \Delta_i,$$

where $A_1, \cdots, A_r$ are formulas of one of the types $(b_0), (b_1), (b_2)$ or (b_3), and $\Gamma_i', \Gamma_i'' \to \Delta_i$ is a prime sequent. Let us denote the deduction terminating with the occurrence S_i by D_i. Let N be the canonical partition for the sequent $\Gamma_i', A_1, \cdots, A_r, \Gamma_i'' \to \Delta_i$. With the help of Theorem 3.1 of [13] we construct for each $i = 1, \cdots, n$ a deduction of the sequent $\Gamma_i', A_1, \cdots, A_r, \Gamma_i'' \to \Delta_i$ in K_0 which is consistent with the partition N. With the help of Lemma 15 we construct the deduction D'_i of the sequent $\Gamma'_i, \Gamma''_i \to \Delta_i$ $(i = 1, \cdots, n)$ in J. We eliminate from each D'_i $(i = 1, \cdots, n)$ all occurrences of instances and quasi-instances of formulas arising from explicitly shown occurrences of formulas of $\overline{\mathrm{E}}\mathrm{q}$ $(=, P_1, \cdots, P_l, f_1, \cdots, f_k)$ in (4).

Let us denote by D''_i the deduction obtained in J for each $i = 1, \cdots, n$. Below $S_1, \cdots, S_n$ in D we cancel all occurrences of formulas which are ancestors of explicitly shown occurrences of the formulas of $\overline{\mathrm{E}}\mathrm{q}$ $(=, P_1, \cdots, P_l, f_1, \cdots, f_k)$ in (4), eliminate repetitions of sequents, and replace each deduction D_i by D''_i. The configuration obtained will be the desired deduction of the sequent $\Gamma \to \Delta$ in $K_0^=$. The lemma is proved.

LEMMA 17. *Let $P_1, \cdots, P_l, f_1, \cdots, f_k$ be the complete list of predicate and function symbols occurring in the sequent $\Gamma \to \Delta$. The sequent $\Gamma \to \Delta$ is deducible in $K_{0c}^=$ if and only if sequent* (4) *is deducible in K_{0c}.*

PROOF. If the sequent $\Gamma \to \Delta$ is deducible in $K_{0c}^=$, sequent (4) is deducible in the classical predicate calculus K_{0c} (cf. Theorem 41 of [14]).

Since all formulas of the list $\overline{\mathrm{E}}\mathrm{q}$ $(=, P_1, \cdots, P_l, f_1, \cdots, f_k)$ are deducible in $K_{0c}^=$, a deduction of the sequent $\Gamma \to \Delta$ in $K_{0c}^=$ can be obtained from a deduction of (4) in K_{0c} with the help of several applications of the cut rule.

Lemmas 16, 17, and the equiextensionality of K_{0c} and K_0 imply the following assertion.

THEOREM 1. *$K_0^=$ and $K_{0c}^=$ are equiextensional.*

§4. Permutability of applications of rules in $K_0^=$

The concepts of a regular partition of a system of classes of occurrences of logical symbols for a sequent, of the consistency of a deduction with a partition, and of the classical almost consistency of a deduction with a partition (cf. [13]) carry over in a natural way to the calculus $K_0^=$.

THEOREM 2. *Whatever the sequent S, the regular partition $\mathfrak{C}$ for S and the regular deduction D of S in the calculus $K_0^=$, any of whose axioms is an axiom over elementary formulas, if the deduction D is almost consistent with $\mathfrak{C}$, then D can be reconstructed into a regular deduction of S in $K_0^=$ which is consistent with $\mathfrak{C}$ and in which all axioms are axioms over elementary formulas.*

PROOF. If the sequent $\Gamma \to \Delta$ is regularly deducible in $K_0^=$, then the sequent $\overline{\mathrm{Eq}}$ $(=, P_1, \cdots, P_l, f_1, \cdots, f_k)$, $\Gamma \to \Delta$ is deducible in K_0 $(P_1, \cdots, P_l, f_1, \cdots, f_k$ is the complete list of predicate and function symbols occurring in $\Gamma \to \Delta)$ (cf. Theorem 41 of [14]). Let $\mathfrak{C}$ be the classically regular partition for the sequent $\Gamma \to \Delta$. With the help of Lemma 16 it is possible to construct a regular deduction of $\Gamma \to \Delta$ in $K_0^=$ which is consistent with $\mathfrak{C}$.

§5. The calculus $L_c^=$

Let L be some applied predicate calculus which has as its apparatus of logical inference the logical tools of the calculus $K_{0c}^=$.

Let us call closed formulas of the form

$$\forall x_1 \ldots x_n (t = w), \tag{1}$$

where t and w are quasi-terms, *formulas of type* I. Let us put into correspondence with each specific axiom of the calculus L which is a formula of type I determinative rules which we shall call the rules associated with formula (1).

Let us call the rules

$$\frac{\Gamma \to [\Delta|_{\tilde{t}}^{\alpha}] \quad \alpha,\ \Gamma,\ \Delta \text{ satisfy Condition } (*)}{\Gamma \to [\Delta|_{\tilde{w}}^{\alpha}]}, \tag{1a}$$

$$\frac{\Gamma \to [\Delta|_{\tilde{w}}^{\alpha}] \quad \alpha,\ \Gamma,\ \Delta \text{ satisfy Condition } (*)}{\Gamma \to [\Delta|_{\tilde{t}}^{\alpha}]}, \tag{1b}$$

the *rules associated with formula* (1), where α is a free variable, Γ and Δ are lists of formulas, $T_1, \cdots, T_n$ are any terms, $\tilde{t}$ denotes $[t|_{T_1,\ldots,T_n}^{x_1,\ldots,x_n}]$, and $\tilde{w}$ denotes $[w|_{T_1,\ldots,T_n}^{x_1,\ldots,x_n}]$.

CONDITION (*). 1) $\Gamma \to \Delta$ is a prime sequent. 2) α occurs in only one member of the list of formulas Δ, and that member is an elementary formula.

If all specific axioms of the calculus L are formulas of type I, we say that L is a calculus of type I. Many equational algebraic theories may be presented as calculi of type I, i.e. theories all of whose nonlogical axioms are equalities, for example, elementary group theory, elementary abelian group theory or elementary ring theory.

Let us call closed formulas of the form

$$\forall x_1 \ldots x_m M. \tag{2}$$

formulas of type II, where M is a quasi-formula which does not contain quantifiers and is not an equality. If each specific axiom of the calculus L is a formula of type I or type II, we call L a calculus of type II. Any applied predicate calculus can be brought to a calculus of type II by reduction of its specific axioms to prenex form and successive elimination of existential quantifiers with the help of the Skolem method (cf. [19]).

Let us call the sequent $\Gamma \to \Delta$ a *generalized prime sequent* if each member of the list Γ is either a quantifier-free formula or a formula beginning with $\forall$, while each member of the list Δ is either a quantifier-free formula or a formula beginning with $\exists$.

Let us call the rule

$$\frac{\left[M \left|{}^{x_1, \ldots, x_m}_{T_1, \ldots, T_m}\right.\right],\ \Gamma \rightarrow \Delta \qquad \Gamma,\ \Delta,\ T_1,\ \ldots,\ T_m \text{ satisfy Condition } (**)}{\Gamma \rightarrow \Delta} \tag{3}$$

the *rule associated with formula* (2).

CONDITION (**). $\Gamma \rightarrow \Delta$ is a generalized primary sequent.

Let L be an arbitrary calculus of type II. We denote by $L^{=}$ the calculus obtained from $K_0^{=}$ in the following way: with respect to each specific axiom of L, the rules associated with it are constructed and adjoined to $K_0^{=}$. We denote by $L_c^{=}$ the calculus obtained from $L^{=}$ by adding all structural rules.

LEMMA 18. *The calculi L and $L_c^{=}$ are equiextensional.*

PROOF. All specific axioms of L are deducible in $L_c^{=}$. Rules (1a), (1b) and (3) of $L_c^{=}$ are deducible in L.

§6. Elimination of structural rules from $L_c^{=}$

It is easy to prove the following assertion.

LEMMA 19. *Let the sequents S and T differ from each other only by a permutation of terms. Then with respect to every deduction of S in $K_0^{=}$, by permuting some terms of the sequents occurring in a given deduction, it is possible to construct a deduction of T in $K_0^{=}$.*

This lemma implies that permutation rules are admitted in $K_0^{=}$.

Let us denote by $\widetilde{L}^{=}$ the calculus obtained from $L^{=}$ by removing all quantifier rules, and by L_0 the calculus obtained from J by adding rules (1a) and (1b).

LEMMA 20. *Let $A_1, \cdots, A_n, B_1, \cdots, B_m$ be specific axioms of L in which $A_1, \cdots, A_n$ are formulas of type* I *and $B_1, \cdots, B_m$ are formulas of type* II. *Let a deduction D of the sequent $\Gamma, A_1, \cdots, A_n, B_1, \cdots, B_m, \Theta \rightarrow \Delta$ in $K_0^{=}$ be given in which all applications of rules are applications of rules pertaining to the explicitly shown occurrences of the formulas $A_1, \cdots, A_n, B_1, \cdots, B_m$ in the sequent $\Gamma, A_1, \cdots, A_n, B_1, \cdots, B_m, \Theta \rightarrow \Delta$ or are applications of rules for equality, while all axioms are axioms over elementary formulas. Then it is possible to construct a deduction of the sequent $\Gamma, \Theta \rightarrow \Delta$ in $\widetilde{L}^{=}$.*

PROOF. The conditions of the lemma imply that $\Gamma, \Theta \rightarrow \Delta$ is a prime sequent.

Let us divide the occurrences of logical symbols in $\Gamma, A_1, \cdots, A_n, B_1, \cdots, B_m, \Theta \rightarrow \Delta$ into two classes, C_1 and C_2. (We shall say that C_i is "higher" than C_j if $i < j$.) Let us attribute to the first class all occurrences of propositional connectives in $B_1, \cdots, B_m$, and to the second class all occurrences of the logical symbol $\forall$ in $A_1, \cdots, A_n, B_1, \cdots, B_m$. This regular partition for the sequent $\Gamma, A_1, \cdots, A_n, B_1, \cdots, B_m, \Theta \rightarrow \Delta$ we shall denote by $\mathfrak{C}$.

With the help of Theorem 2 we shall reconstruct D into a regular deduction D_1 consistent with $\mathfrak{C}$.

We denote by S the sequent which is the conclusion of the lowest application of the rules of C_1 in D_1. The sequent S has (up to a permutation of terms) the form $\Gamma, E, \Lambda, \Pi, \Sigma, \Theta \to \Delta$, where E is a list of instances of formulas $A_1, \cdots, A_n$, Λ is a list consisting of formulas $A_1, \cdots, A_n$ and quasi-instances of them, Π is a list of instances of formulas $B_1, \cdots, B_m$, and Σ is a list consisting of formulas $B_1, \cdots, B_m$ and quasi-instances of them.

We call the explicitly shown occurrence of the formula $[M|^{x_1,\dots,x_m}_{T_1,\dots,T_m}]$ in the premise of rule (3) the *principal formula* of rule (3). The principal formula is always an instance of a formula of type II.

Let us write the list Π as $M'_1, \cdots, M'_r$. Each of the formulas M'_k can be the principal formula of rule (3).

It is easy to see that the sequent $\Gamma, E, \Pi, \Theta \to \Delta$ is deducible in $K_0^=$. Let us denote by D_0 a deduction of this sequent in $K_0^=$. All applications of rules in D_0 pertain to C_1 or are applications of rules for equality. By virtue of Lemma 19, the sequent $M'_1, \cdots, M'_r, \Gamma, E_1, \Theta \to \Delta$ is deducible in $K_0^=$, with all applications of rules in a deduction of this sequent pertaining to explicitly shown occurrences of formulas $M'_1, \cdots, M'_r$ in the sequent $M'_1, \cdots, M'_r, \Gamma, E, \Theta \to \Delta$ or being applications of rules for equality.

With the help of Lemmas 1 and 16, let us construct a regular deduction D_2 of the sequent $M'_1, \cdots, M'_r, \Gamma, E, \Theta \to \Delta$ in $K_0^=$ in which all applications of rules for equality are succedent and all axioms are axioms over elementary formulas. Let us replace each axiom of the form $\Gamma'_1, r = s, \Gamma'_2 \to \Delta'_1, r = s, \Delta'_2$ in D_2 by the configuration

$$\frac{\Gamma'_1,\ r = s,\ \Gamma'_2 \to \Delta'_1,\ r = r,\ \Delta'_2}{\Gamma'_1,\ r = s,\ \Gamma'_2 \to \Delta'_1,\ r = s,\ \Delta'_2} \to E.$$

Let us denote the new deduction by D_3. D_3 coincides with D_2 if there is not an axiom of the form $\Gamma'_1, r = s, \Gamma'_2 \to \Delta'_1, r = s, \Delta'_2$ in D_2.

Then in D_3 let us replace all applications of the rules $\to E$ whose auxiliary formulas are ancestors of the explicitly shown occurrences of the formulas E in the sequent $M'_1, \cdots, M'_r, \Gamma, E, \Theta \to \Delta$, respectively by applications of the rules (1a) and (1b).

Next let us cancel in the configuration obtained all occurrences of formulas which are ancestors of explicitly shown occurrences of E in the sequent $M'_1, \cdots, M'_r, \Gamma, E, \Theta \to \Delta$. The configuration obtained will be a deduction of the sequent $M'_1, \cdots, M'_r, \Gamma, \Theta \to \Delta$ in $\widetilde{L}^=$.

Let us denote by D_4 this deduction of the sequent $M'_1, \cdots, M'_r, \Gamma, \Theta \to \Delta$ in $\widetilde{L}^=$.

Hence the configuration

$$\dfrac{D_4\left\{\begin{array}{c}\cdots\\ M'_1,\ M'_2,\ \ldots,\ M'_{r-1},\ M'_r,\ \Gamma,\ \Theta\to\Delta\end{array}\right.}{M'_2,\ M'_3,\ \ldots,\ M'_{r-1},\ M'_r,\ \Gamma,\ \Theta\to\Delta}\ (3)$$

$$\dfrac{\dfrac{\dfrac{\cdots}{M'_{r-1},\ M'_r,\ \Gamma,\ \Theta\to\Delta}}{M'_r,\ \Gamma,\ \Theta\to\Delta}\ (3)}{\Gamma,\ \Theta\to\Delta}\ (3).$$

will be a deduction of the sequent $\Gamma, \Theta \to \Delta$ in $\widetilde{L}^{=}$. The lemma is proved.

LEMMA 21. *If the sequent $\Gamma \to \Delta$ is deducible in $L_0^{=}$, it is possible to find a list $C_1, \cdots, C_l$ of specific axioms of the calculus L such that the sequent $C_1, \cdots, C_l$, $\Gamma \to \Delta$ is deducible in $K_{0c}^{=}$.*

PROOF. Let a deduction D of the sequent $\Gamma \to \Delta$ in $L_c^{=}$ be given. Then it is possible to find a list $C_1, \cdots, C_l$ of specific axioms of L such that $C_1, \cdots, C_l, \Gamma \to \Delta$ is deducible in $K_{0c}^{=}$. Indeed, let us put the formula $\forall x_1 \cdots x_n\ (t = w)$ into correspondence with each application of the rule (1a) or (1b) and put the formula $x_1 \cdots x_m M$ into correspondence with each application of rule (3).

Let $A_1, \cdots, A_n$ be the complete list of formulas put into correspondence with the applications of rules (1a) and (1b) in D, and let $B_1, \cdots, B_m$ be the complete list of formulas put into correspondence with the applications of rules (3) in D. It is obvious that $A_1, \cdots, A_n$ are formulas of type I and $B_1, \cdots, B_m$ are formulas of type II.

We ascribe to the antecedent of each sequent encountered in D the formulas $A_1, \cdots, A_n, B_1, \cdots, B_m$. Each application of a rule remains (up to a permutation of formulas) an application of the same rule. Let us denote the configuration obtained by D'.

With the help of structural rules, the rules for equality, $\to \exists$ and $\forall \to$, it is easy to write insertions for those applications of rules in D' which differ from applications of rules of $K_{0c}^{=}$ so that they turn into applications of rules of $K_{0c}^{=}$.

Thus we obtain a deduction of the sequent $A_1, \cdots, A_n, B_1, \cdots, B_m, \Gamma \to \Delta$ in $K_{0c}^{=}$.

LEMMA 22. *If it is possible to find a list $C_1, \cdots, C_l$ of specific axioms of the calculus L such that the sequent $C_1, \cdots, C_l, \Gamma \to \Delta$ is deducible in $K_0^{=}$, then the sequent $\Gamma \to \Delta$ is deducible in $L^{=}$.*

PROOF. Let a deduction of $C_1, \cdots, C_l, \Gamma \to \Delta$ in $K_0^{=}$ be given. With the help of Lemma 16 and Theorem 2 we construct a regular deduction D of $C_1, \cdots, C_l$, $\Gamma \to \Delta$ in $K_0^{=}$ in which all applications of rules pertaining to explicitly shown occurrences of formulas $C_1, \cdots, C_l$ in the sequent $C_1, \cdots, C_l, \Gamma \to \Delta$, are higher than all applications of rules pertaining to formulas of Γ and Δ, and all axioms are axioms over elementary formulas.

Let us write in the form $S_1, \cdots, S_r$ the complete list of occurrences of the prime sequents in D which are conclusions of the lowest applications of rules pertaining to explicitly shown occurrences of the formulas $C_1, \cdots, C_l$ in the sequent $C_1, \cdots, C_l$, $\Gamma \to \Delta$. For each $i = 1, \cdots, r$ the basis of S_i has the form $\Gamma_i', C_1, \cdots, C_l, \Gamma_i'' \to \Delta_i$.

Let us denote by D_i' a deduction ending with an occurrence of S_i. With the help of Lemma 20, for each $i = 1, \cdots, r$ we construct a deduction D_i'' of $\Gamma_i', \Gamma_i'' \to \Delta_i$ in $\widetilde{L}^=$.

In the deduction D_2 let us replace each deduction D_i' by D_i'' and cancel below each occurrence of the sequent $\Gamma_i', \Gamma_i'' \to \Delta_i$ all occurrences of formulas which are ancestors of explicitly shown occurrences of the formulas $C_1, \cdots, C_l$ in the sequent $C_1, \cdots, C_l, \Gamma \to \Delta$. The configuration obtained will be a deduction of $\Gamma \to \Delta$ in $L^=$. The lemma is proved.

Lemmas 21, 22 and the equiextensionality of $K_{0c}^=$ and $K_0^=$ imply the following assertion.

THEOREM 3. *$L_c^=$ and $L^=$ are equiextensional.*

Let us denote by $\overline{K_0^=}$ (by $\overline{L^=}$) the calculus obtained from $K_0^=$ (from $L^=$) by replacing rules (1) for equality by the two rules for equality:

$$\frac{\Gamma_1,\ r=s,\ \Gamma_2 \to \Delta_1,\ [E\,|^{\alpha}_{r}],\ \Delta_2}{\Gamma_1,\ r=s,\ \Gamma_2 \to \Delta_1,\ [E\,|^{\alpha}_{s}],\ \Delta_2},\quad \frac{\Gamma_1,\ s=r,\ \Gamma_2 \to \Delta_1,\ [E\,|^{\alpha}_{r}],\ \Delta_2}{\Gamma_1,\ s=r,\ \Gamma_2 \to \Delta_1,\ [E\,|^{\alpha}_{s}],\ \Delta_2}, \tag{1'}$$

where $\Gamma_1, r = s, \Gamma_2 \to \Delta_1, E, \Delta_2$ is a prime sequent and E is an elementary formula.

Theorems 1 and 3 and Lemmas 1, 3 and 16 imply that $\overline{K_0^=}$ and $K_0^=$ ($\overline{L^=}$ and $L^=$) are equiextensional.

EXAMPLE. The construction cited above for forming the calculus $L^=$ permits us to present, for example, elementary group theory as the calculus $\mathfrak{I}$ obtained from $K_0^=$ by adding the rules for replacing the terms $f(e, T)$ by T, $f(f(t_1, t_2), t_3)$ by $f(t_1, f(t_2, t_3))$ and $f(r(t), t)$ by e, as well as the replacement rules inverse to these, where T, t, t_1, t_2, t_3 are terms, e is the group identity element, $r(x_1)$ is the element inverse to x_1 and $f(x_1, x_2)$ is the product of x_1 and x_2.

In rule (3) of §5 it is possible, in addition, to introduce the following restriction: the premise immediately splits up to prime sequents.

Let M be a quantifier-free formula of $L^=$ which is not an equality of two terms. It is always possible to find a list of elementary sequents $\Theta_1 \to \Pi_1, \cdots, \Theta_n \to \Pi_n$, such that the sequent $M \to$ is deducible from the sequents $\Theta_i \to \Pi_i$ $(i = 1, \cdots, n)$ with the help of the propositional rules of $L^=$. (Θ_i and Π_i are nonempty lists of elementary formulas.) In the case when M is an elementary formula, $\Theta_1 \to \Pi_1 \stackrel{\circ}{=} M \to$. $\Theta_i \to \Pi_i$ can be found uniquely by splitting $M \to$ up to elementary sequents with the help of the propositional rules of $L^=$ (passing from conclusion to premises).

Let us consider a formula $\forall x_1 \cdots x_m M$ of type II which is a specific axiom of $L^=$.

Let $[M|_{T_1,\dots,T_m}^{x_1,\dots,x_m}]$ be an instance of this formula.

Let us denote by $R(\Gamma \to \Delta;\ \forall x_1 \cdots x_m M;\ T_1, \cdots, T_m)$ the list of sequents

$$\left[\Theta_1 |_{T_1,\dots,T_m}^{x_1,\dots,x_m}\right],\ \Gamma \to \Delta,\ \left[\Pi_1 |_{T_1,\dots,T_m}^{x_1,\dots,x_m}\right],\ \dots,\ \left[\Theta_k |_{T_1,\dots,T_m}^{x_1,\dots,x_m}\right],\ \Gamma \to \Delta,\ \left[\Pi_k |_{T_1,\dots,T_m}^{x_1,\dots,x_m}\right].$$

We recall that $[\Theta|_{T_1,\dots,T_m}^{x_1,\dots,x_m}]$ denotes the result of simultaneously substituting the terms $T_1, \cdots, T_m$, respectively, in place of $x_1, \cdots, x_m$ in the list of formulas Θ.

Let us put a determinative rule into correspondence with each specific axiom of L which is a formula of type II. We call the rule

$$\frac{\begin{array}{c} R(\Gamma \to \Delta;\ \forall x_1 \dots x_m M;\ T_1. \ \dots,\ T_m) \\ \Gamma,\ \Delta,\ T_1,\ \dots,\ T_m \quad \text{satisfy Condition } (***) \end{array}}{\Gamma \to \Delta} \tag{3'}$$

the rule associated with the formula $\forall x_1 \cdots x_m M$.

CONDITION (***). $\Gamma \to \Delta$ is a prime sequent.

Let us denote by $\tilde{\tilde{L}}^=$ the calculus obtained from $L^=$ by replacing rule (3) with rule (3′), and by $\tilde{\tilde{L}}_c^=$ the calculus obtained from $\tilde{\tilde{L}}^=$ by adding the two structural rules.

The following assertion is proved analogously to Theorem 3.

THEOREM 4. $\tilde{\tilde{L}}_c^=$ *and* $\tilde{\tilde{L}}^=$ *are equiextensional.*

§7. Sequential variant of the calculus of equivalence

A. Janiczak in [20] considered the decidable calculus T_3. The formulas of this calculus are constructed in the usual way from object variables, a 2-place predicate symbol R_0 and equality $=$ with the help of logical connectives. The calculus T_3 has three specific axioms:

1) $\forall x_1 (x_1 R_0 x_1)$,
2) $\forall x_1 x_2 (x_1 R_0 x_2 \supset x_2 R_0 x_1)$,
3) $\forall x_1 x_2 x_3 ((x_1 R_0 x_2 \,\&\, x_2 R_0 x_3) \supset x_1 R_0 x_3)$

and, as the apparatus of logical inference, the logical tools of the classical predicate calculus with equality, without function symbols.

Let us formulate this apparatus in the form of the calculus $Я_c^=$, which differs from $K_{0c}^=$ only in that the formulas of $Я_c^=$ are constructed exactly as the formulas of T_3. Let us denote by $T_{3c}^=$ the calculus obtained from $Я_c^=$ by adding the schema of axioms $\Gamma \to \Delta_1,\ tR_0 t,\ \Delta_2$ and the rules

$$\frac{[\Gamma |_r^\alpha],\ rR_0 s,\ [\Delta |_r^\alpha] \to [\Theta |_r^\alpha]}{[\Gamma |_s^\alpha],\ rR_0 s,\ [\Delta |_s^\alpha] \to [\Theta |_s^\alpha]},\quad \frac{[\Gamma |_r^\alpha],\ sR_0 r,\ [\Delta |_r^\alpha] \to [\Theta |_r^\alpha]}{[\Gamma |_s^\alpha],\ sR_0 r,\ [\Delta |_s^\alpha] \to [\Theta |_s^\alpha]}, \tag{1}$$

where α is an arbitrary free variable occurring in only one of the members of the list of formulas Γ, Δ, Θ, which is an *elementary formula* of the form $uR_0 w$ (u and w are arbitrary terms).

It is easy to prove the equiextensionality of T_3 and $T^{=}_{3c}$.

Let us denote by $T^{=}_3$ the calculus obtained from $T^{=}_{3c}$ by eliminating all structural rules. The equiextensionality of $T^{=}_{3c}$ and $T^{=}_3$ is proved analogously to Theorem 1.

§8. Sequential variants of the calculi of R. Robinson and M. Presburger

Let us consider the arithmetical system of Raphael M. Robinson [14], which is well known to be essentially undecidable although finitely axiomatizable. Let us denote Robinson's calculus by R_0. As we know, the specific axioms of R_0 are, besides recursive equations for addition and multiplication, the axioms

1) $\forall x_1 x_2 (x_1' = x_2' \supset x_1 = x_2)$,
2) $\forall x_1 \neg (x_1' = 0)$,
3) $\forall x_1 (x_1 = 0 \vee \exists x_2 (x_1 = x_2'))$.

Let us introduce into the discussion the calculus R_1 obtained from R_0 by replacing axiom 3) with the axiom

3_1) $\forall x_1 (x_1 = 0 \vee x_1 = (\mathrm{Pd}\ (x_1))')$,

where Pd is a new one-place function symbol. ($\mathrm{Pd}(x_1)$ is read "predecessor of x_1".) Formula 3_1) is obtained from 3) by applying the Skolem method of eliminating quantifiers. Therefore R_1 is an unessential extension of R_0, i.e. a formula not containing the symbol Pd is deducible in R_0 if and only if it is deducible in R_1 [19].

Let us denote by R_2 the calculus obtained from R_1 by replacing axiom 1) with the axiom

1_1) $\forall x_1 (x_1 = \mathrm{Pd}\ (x_1'))$.

R_2 is a calculus of type II. It is easy to prove the equiextensionality of R_1 and R_2.

The construction cited above for forming $\widetilde{\widetilde{L}}^{=}$ permits us to prove the equiextensionality of R_2 and the calculus R_3 obtained from $K^{=}_0$ by adding the rules associated with the specific axioms of type I of R_2 and the rules

$$\frac{\Gamma \to \Delta \qquad \begin{array}{c}\Gamma \to \Delta,\ t' = 0 \\ \text{is a prime sequent}\end{array}}{\Gamma \to \Delta}$$

$$\frac{\begin{array}{c} t = 0,\ \Gamma \to \Delta;\ t = (\mathrm{Pd}\ (t))',\ \Gamma \to \Delta \\ \Gamma \to \Delta \text{ is a prime sequent}\end{array}}{\Gamma \to \Delta}$$

M. Presburger [21] considered a decidable calculus whose specific axioms are formulas expressing the commutativity and associativity of addition, the axioms

1) $\forall x_1 (x_1 + 0 = x_1)$,
2) $\forall x_1 x_2 \exists x_3 (x_1 + x_3 = x_2)$

and the three formula schemata

3) $\forall x_1 x_2 (\alpha x_1 = \alpha x_2 \supset x_1 = x_2)$, ($\alpha x$ denotes $\underbrace{x + x + \ldots + x}_{\alpha \text{ times}}$, $\alpha = 2, 3, \ldots$),

4) $\forall x_1 \exists x_2 [(\alpha x_2 = x_1) \vee (\alpha x_2 + 1 = x_1) \vee \ldots \vee (\alpha x_2 + \underbrace{1 + \ldots + 1}_{(\alpha - 1) \text{ times}} = x_1)]$ $(\alpha = 2, 3, \ldots)$,

5) $\forall x_1 \neg (\alpha x_1 + 1 = 0)$ $(\alpha = 2, 3, \ldots)$.

Let us denote this Presburger calculus by Π_0.

Integers are the admissible values of the object variables in this system. Formula 2) is a symbolization of an assertion that the difference of two numbers always exists; formula 4) is a symbolization of an assertion that for each natural number α and for each integer x_1 an integer x_2 exists such that $\alpha x_2 + i = x_1$, where i is a natural number such that $i < \alpha$, and formula 5) is a symbolization of an assertion that the equation $\alpha a + 1 = 0$ does not have integral solutions for $\alpha > 2$.

Let us introduce into the discussion the calculus Π_1 obtained from Π_0 by replacing axiom 2) with the axiom

2_1) $\forall x_1 x_2 (x_1 + \varphi(x_1, x_2) = x_2)$

and axiom schema 4) with the axiom schema

4_1) $\forall x_1 [(\alpha g_\alpha(x_1) = x_1) \vee (\alpha g_\alpha(x_1) + 1 = x_1) \vee \ldots \vee (\alpha g_\alpha(x_1) + \underbrace{1 + \ldots + 1}_{(\alpha - 1) \text{ times}} = x_1)]$,

where φ is a new two-place function symbol and, for each $\alpha = 2, 3, \cdots$, g_α is a new one-place function symbol.

Formula 2_1) and axiom schema 4_1) are obtained, respectively, from 2) and 4) by applying the Skolem method of eliminating quantifiers. Therefore Π_1 is an unessential extension of Π_0, i.e. a formula not containing the symbols φ and g_α $(\alpha = 2, 3, \cdots)$ is deducible in Π_0 if and only if it is deducible in Π_1. The calculus Π_1 is a calculus of type II.

The construction cited above for forming $\tilde{\tilde{L}}^=$ permits us to prove the equiextensionality of Π_1 and the calculus Π_2 obtained from $K_0^=$ by adding the rules associated with the specific axioms of type I of Π_1, as well as the rules

$$\frac{\Gamma \to \Delta,\ ar = as;\ r = s,\ \Gamma \to \Delta \quad \Gamma \to \Delta \text{ is a prime sequent and } \alpha = 2, 3, \ldots}{\Gamma \to \Delta},$$

$$a g_\alpha(t) = t,\ \Gamma \to \Delta;\ \ a g_\alpha(t) + 1 = t,\ \Gamma \to \Delta;\ \ldots;\ a g_\alpha(t) + \underbrace{1 + \ldots + 1}_{(\alpha - 1) \text{ times}} = t,\ \Gamma \to \Delta$$

$$\frac{\Gamma \to \Delta \text{ is a prime sequent and } \alpha = 2, 3, \ldots}{\Gamma \to \Delta},$$

$$\frac{\begin{array}{c}\Gamma \to \Delta,\ \alpha T + 1 = 0\\ \Gamma \to \Delta \text{ is a prime sequent and } \alpha = 2, 3, \ldots\end{array}}{\Gamma \to \Delta}.$$

§9. Some sequential variants of applied predicate calculi

Let us call closed formulas of the form

$$\forall x_1 \ldots x_m P \tag{1}$$

formulas of type I^+ (cf. [22]), where P is an elementary quasi-formula which is not an equality of two quasi-terms.

Let us call closed formulas of the form

$$\forall x_1 \ldots x_l \neg Q, \tag{2}$$

formulas of type I^-, where Q is an elementary quasi-formula which is not an equality of two quasi-terms.

Let us call closed formulas of the form

$$\forall x_1 \ldots x_n (t = w), \tag{3}$$

formulas of type $\mathrm{I}^=$, where t and w are quasi-terms.

Let us call closed formulas not attributable to any of the types $\mathrm{I}^+, \mathrm{I}^-, \mathrm{I}^=$ and of the form

$$\forall x_1 \ldots x_s M, \tag{4}$$

formulas of type II', where M is a quantifier-free quasi-formula.

Let L' be some applied predicate calculus having as apparatus of logical deduction the logical tools of the calculus $K_0^=$. If each specific axiom of L' is a formula of one of the types $\mathrm{I}^+, \mathrm{I}^-, \mathrm{I}^=$ or II', where the predicate symbols occurring in axioms of type I^+ differ from the predicate symbols occurring in axioms of type I^-, let us call L' a *calculus of type* II'.

We put determinative axiom schemata and rules into correspondence with each specific axiom of L'. We call the schema

$$\Gamma \to \Delta_1,\ \widetilde{P},\ \Delta_2, \tag{1a}$$

an axiom schema *associated with formula* (1), where $\widetilde{P}$ denotes $[P|^{x_1,\ldots,x_m}_{T_1,\ldots,T_m}]$ and $T_1, \cdots, T_m$ are terms.

Let us call the schema

$$\Gamma_1,\ \widetilde{Q},\ \Gamma_2 \to \Delta, \tag{2a}$$

an axiom schema *associated with formula* (2), where $\widetilde{Q}$ denotes $[Q|^{x_1,\ldots,x_l}_{s_1,\ldots,s_l}]$ and $s_1, \cdots, s_l$ are terms.

Let us call the rule

$$\frac{\alpha,\ \Gamma,\ \Delta\quad \dfrac{\left[\Gamma\,|_{\tilde{t}}^{\alpha}\right]\to\left[\Delta\,|_{\tilde{t}}^{\alpha}\right]}{\text{satisfy Condition }(*)'}}{\left[\Gamma\,|_{\tilde{w}}^{\alpha}\right]\to\left[\Delta\,|_{\tilde{w}}^{\alpha}\right]} \tag{3a}$$

and the inverse rule, rules *associated with formula* (3); here α is a free variable, $r_1, \cdots, r_n$ are terms, $\tilde{t}$ denotes $[t|_{r_1,\dots,r_n}^{x_1,\dots,x_n}]$, and $\tilde{w}$ denotes $[w|_{r_1,\dots,r_n}^{x_1,\dots,x_n}]$.

CONDITION $(*)'$. 1) $\Gamma \to \Delta$ is a prime sequent. 2) α occurs in only one member of the list of formulas Γ, Δ, and this member is an elementary formula.

The rule associated with formula (4) is determined as in §5.

Let us denote by J_1 the auxiliary calculus obtained from J by adding all schemata associated with formulas (1) and (2) and all rules associated with formula (3). Let us denote by Θ^+ an arbitrary list of instances of formulas of type I^+, by Θ^* an arbitrary list of instances of formulas of type I^-, by Θ^- the results of canceling negations from Θ^*, and by $\Theta^=$ an arbitrary list of instances of formulas of type $\mathrm{I}^=$.

LEMMA 23. *It is possible to construct relative to any deduction D of the sequent $\Theta^+, \Theta^=, \Gamma \to \Delta, \Theta^-$ in J a deduction of the sequent $\Gamma \to \Delta$ in J_1.*

PROOF. With the help of Lemma 4, let us reconstruct D into a succedent weeded deduction D_1 in J. A number of cases are possible here, of which the following are the most important:

1) No occurrence of Θ^+, $\Theta^=$ or Θ^- is traceable to an axiom.

2) The only occurrence of Θ^+ is traceable to an axiom and no occurrence of Θ^- is traceable to an axiom.

3) The only occurrence of Θ^- is traceable to an axiom and no occurrence of Θ^+ is traceable to an axiom.

Let us examine case 3). We write the lists Θ^- and Γ in the form $\Theta_1^-, \tilde{\Theta}, \Theta_2^-$ and $\Gamma_1, \Theta', \Gamma_2$, respectively. Then the highest sequent of D_1 can be written as

$$\Theta^+,\ \Theta^=,\ \Gamma_1,\ \Theta',\ \Gamma_2 \to \Delta,\ \Theta_1^-,\ \Theta',\ \Theta_2^-.$$

With the help of Lemma 6 we construct a weeded deduction D_2 of the sequent $\Theta^+, \Theta^=, \Gamma_1, \tilde{\Theta}, \Gamma_2 \to \Delta, \Theta_1^-, \Theta', \Theta_2^-$ in J in which $\Theta^+, \Theta^=, \Gamma_1, \tilde{\Theta}, \Gamma_2 \to \Delta, \Theta_1^-, \tilde{\Theta}, \Theta_2^-$ is the topmost sequent. Next we replace in D_2 all antecedent applications of rules for equality whose auxiliary formulas are ancestors of explicitly shown occurrences of formulas of Θ by corresponding applications of rule (3a) and the rule inverse to it. Then in the configuration obtained we cancel all occurrences of formulas which are ancestors of explicitly shown occurrences of formulas of Θ^+, Θ^-, $\Theta^=$ and all repetitions of sequents. The configuration obtained will be an antecedent deduction of the sequent $\Gamma \to \Delta$ in J_1, since its upper sequent is an axiom of J_1.

The remaining cases are considered analogously.

Let L' be an arbitrary calculus of type II$'$. We denote by $L'^{=}$ the calculus obtained from $K_0^{=}$ in the following way: 1) With respect to each specific axiom of L' which is a formula of type $\mathrm{I}^{=}$ or type II$'$, the rules associated with them are constructed and adjoined to $K_0^{=}$. 2) With respect to each specific axiom of L' which is a formula of type I^{+} or I^{-}, the axiom schemata associated with it are found and adjoined to $K_0^{=}$.

Let us denote by $L_c'^{=}$ the calculus obtained from $L'^{=}$ by adding all structural rules.

The following assertion is easily proved.

THEOREM 5. *$L_c'^{=}$ and $L'^{=}$ are equiextensional.*

In the sequel the rules $\forall \to$ and $\to \exists$ will be replaced respectively by the rules

$$\forall_1\to,\ \to\exists_1,\ \forall_2\to,\ \to\exists_2,\ \forall_3\to,\ \to\exists_3,\ \forall_4\to,\ \to\exists_4.$$

The corresponding calculi will be indexed accordingly.

Let us denote by $\forall_1 \to$ (by $\to \exists_1$) the rule obtained from the rule $\forall \to$ (from $\to \exists$) of K_0 by imposing the supplementary condition that the characteristic term of the rule is immediately accessible from N and $C(f_1, \cdots, f_l)$, where N is the list of all terms occurring in the conclusion of the rule and $f_1, \cdots, f_l$ is the complete list of function symbols occurring in the conclusion of the rule.

If $C(f_1, \cdots, f_l)$ is empty, the rules $\forall_1 \to$ and $\to \exists_1$ respectively turn into the rules $\forall_0 \to$ and $\to \exists_0$ obtained respectively from $\forall \to$ and $\to \exists$ by imposing the supplementary condition that either the characteristic term occurs in the conclusion of the rule or no terms at all occur in the conclusion and the characteristic term is the variable a_1.

Let us denote by K_1 (by $K_1^{=}$) the calculus obtained from K_0 ($K_0^{=}$) by replacing the rules $\forall \to$ by $\forall_1 \to$ and $\to \exists$ by $\to \exists_1$. Let us denote by K_{1c} (by $K_{1c}^{=}$) the calculus obtained from K_1 (from $K_1^{=}$) by adding all structural rules.

It is easy to prove the following assertions.

THEOREM 6. *K_0 and K_1 ($K_0^{=}$ and $K_1^{=}$) are equiextensional.*

THEOREM 7. *Whatever the sequent S, the regular partition $\mathfrak{C}$ for S and the deduction D of S in K_1, if any axiom of D is an axiom over elementary formulas and D is classically almost consistent with $\mathfrak{C}$, then D can be reconstructed into a deduction D'' of S in K_1 which is consistent with $\mathfrak{C}$.*

The theorem holds, being obtained from Theorem 2 on replacing $K_0^{=}$ by $K_1^{=}$.

Let L be some noncontradictory applied predicate calculus having as apparatus of logical inference the logical tools of the calculus $K_{1c}^{=}$, and let $f_1, f_2, \cdots$ be the set of all function symbols belonging to the language of this calculus. We denote by $C(L)$ the set all quasi-terms $C(f_1), C(f_2), \cdots$.

Let us denote by $\forall_2 \rightarrow$ (by $\rightarrow \exists_2$) the rule obtained from the rule $\forall \rightarrow$ (from $\rightarrow \exists$) of K_0 by imposing the supplementary condition that the characteristic term of the rule is immediately accessible from N and $C(L)$, where N is the set of all terms occurring in the conclusion of the rule or in at least one of the specific axioms of L.

Let us denote by (3″) the rule obtained from rule (3) (cf. §5) of $L^=$ by imposing the second supplementary condition that the terms $T_1, \cdots, T_m$ are immediately accessible from Q and $C(L)$, where Q denotes the set of all terms occurring in at least one of the quantifier-free members of the list Γ, Δ or in at least one of the specific axioms of L.

Let us denote by $L_2^=$ the calculus obtained from $L^=$ by replacing the rules $\forall \rightarrow$, $\rightarrow \exists$ and (3) respectively by the rules $\forall_2 \rightarrow$, $\rightarrow \exists_2$ and (3″). Let us denote by $L_{2c}^=$ the calculus obtained from $L_2^=$ by adding all structural rules.

The following assertion is proved analogously to Theorem 3.

THEOREM 10. *$L_{2c}^=$ and $L_2^=$ are equiextensional.*

Let L be an arbitrary applied predicate calculus of type II, and let $\varphi_1, \varphi_2, \cdots$ be the set of all function symbols occurring in the specific axioms of L. We denote by $C^*(L)$ the set of all quasi-terms $C(\varphi_1), C(\varphi_2), \cdots$.

Let us denote by $\forall_3 \rightarrow$ (by $\rightarrow \exists_3$) the rule obtained from the rule $\forall \rightarrow$ (from $\rightarrow \exists$) of K_0 by imposing the supplementary condition that the characterisitc term of the rule is immediately accessible from N and $C(\psi_1, \cdots, \psi_m) \cup C^*(L)$, where N is the set of all terms occurring in the conclusion of the rule or in at least one of the specific axioms of L and $\psi_1, \cdots, \psi_m$ is the complete list of function symbols occurring in the conclusion of the rule.

Let us denote by (3‴) the rule obtained from rule (3″) of $L_2^=$ on replacing the set of quasi-terms $C(L)$ in the second item of Condition (**) by the set $C(\psi_1, \cdots, \psi_m) \cup C^*(L)$.

Let us denote by $L_3^=$ the calculus obtained from $L_2^=$ by replacing the rules $\forall_2 \rightarrow$, $\rightarrow \exists_2$ and (3″) respectively by the rules $\forall_3 \rightarrow$, $\rightarrow \exists_3$ and (3‴). Let us denote by $L_{3c}^=$ the calculus obtained from $L_3^=$ by adjoining all structural rules.

Let us denote by S_L the set of all quasi-terms occurring in at least one of the specific axioms of the calculus L. Let us denote by $\forall_4 \rightarrow$, $\rightarrow \exists_4$ and (3⁗) the rules obtained respectively from the rules $\forall_3 \rightarrow$, $\rightarrow \exists_3$ and (3‴) of $L_3^=$ in the following way: we replace the set of quasi-terms $C^*(L)$ everywhere by the set S_L in the formulations of the rules $\forall_3 \rightarrow$, $\rightarrow \exists_3$ and (3‴), and we leave the remaining parts of the formulations of these rules unchanged.

Let us denote by $L_4^=$ the calculus obtained from $L_3^=$ on replacing the rules $\forall_3 \rightarrow$, $\rightarrow \exists_3$ and (3‴) respectively by the rules $\forall_4 \rightarrow$, $\rightarrow \exists_4$ and (3⁗). Let us denote by $L_{4c}^=$ the calculus obtained from $L_4^=$ by adjoining all structural rules.

The following two theorems are proved quite analogously to Theorem 3.

THEOREM 11. *$L_{3c}^{=}$ and $L_3^{=}$ are equiextensional.*

THEOREM 12. *$L_{4c}^{=}$ and $L_4^{=}$ are equiextensional.*

If the list of specific axioms of L is empty, then $L_3^{=}$ and $L_4^{=}$ coincide with $K_1^{=}$.

Let us denote by $\overline{L}_3^{=}$ (by $\overline{L}_4^{=}$) the calculus obtained from $L_3^{=}$ (from $L_4^{=}$) by replacing the rules (1) for equality by the rules (1′) for equality. It is easy to see that $\overline{L}_3^{=}$ and $L_3^{=}$ ($\overline{L}_4^{=}$ and $L_4^{=}$) are equiextensional.

In rules (1a) and (1b) of $L^{=}$ let us call the occurrences in the conclusions of formulas in which replacement occurs *principal occurrences of the rules.* Let us call a conclusion's formulas in which replacement occurs *principal formulas.*

Let us denote by (1a′) (by (1b′)) the rule obtained from rule (1a) (from (1b)) of $L^{=}$ by repeating the principal formula of the rule in the premise and appending the third supplementary condition that the terms $T_1, \cdots, T_n$ are immediately accessible from Q' and $C(L)$, where Q' denotes the set of all terms occurring in at least one of the elementary members of the list Γ, Δ or in at least one of the specific axioms of L.

Let us introduce into the discussion the rules for equality

$$\frac{\Gamma_1,\ r=s,\ \Gamma_2 \to [\Delta|^{\alpha}_{s}],\ [\Delta|^{\alpha}_{r}]}{\Gamma_1,\ r=s,\ \Gamma_2 \to [\Delta|^{\alpha}_{s}]}, \qquad \frac{\Gamma_1,\ r=s,\ \Gamma_2 \to [\Delta|^{\alpha}_{r}],\ [\Delta|^{\alpha}_{s}]}{\Gamma_1,\ r=s,\ \Gamma_2 \to [\Delta|^{\alpha}_{r}]}, \qquad (1'')$$

where α is an arbitrary free variable occurring in only one of the members of the list of formulas Δ, it being an elementary formula; $\Gamma_1, r=s, \Gamma_2 \to [\Delta|^{\alpha}_{s}]$ is a prime sequent.

We say that the quasi-term t has a surplus relative to the quasi-term w if there is (at least) one bound or free variable occurring in t and not occurring in w. Rules (1a′) and (1b′) are especially convenient to use in searching for a deduction when the quasi-term t (respectively, w) has a surplus relative to the quasi-term w (respectively, t).

Let us denote by $\overline{\overline{L}}_2^{=}$ the calculus obtained from $L_2^{=}$ by replacing rules (1a), (1b) and (1) respectively with rules (1a′), (1b′) and (1″), and by $\overline{\overline{L}}_{2c}^{=}$ the calculus obtained from $\overline{\overline{L}}_2^{=}$ by adding all structural rules.

The following assertion can be proved.

THEOREM 13. *$\overline{\overline{L}}_{2c}^{=}$ and $\overline{\overline{L}}_2^{=}$ are equiextensional.*

REMARK. If in $L_2^{=}$ we replace the rule $\forall_2 \to$ (or $\to \exists_2$) by the rule $\forall_1 \to$ (by $\to \exists_1$), we obtain the calculus L_2^{*}, which is not, generally speaking, equiextensional with $L_2^{=}$.

Indeed, let L be a calculus of type I which has the unique specific axiom

$$\forall x_1 x_2 (\varphi(x_1, f(x_1)) = \varphi(x_2, g(x_1, x_2))),$$

where f is a one-place, and φ and g are two-place, function symbols.

The sequent $\to \exists x_1 x_2(\varphi(a_1, x_1) = \varphi(a_2, x_2))$ (a_1 and a_2 are distinct free variables), for example, which is deducible in $L_2^=$, is not deducible in L_2^*. In fact, the following configuration, for example,

$$\dfrac{\dfrac{\begin{array}{c}\to \varphi(a_1, f(a_1)) = \varphi(a_1, f(a_1)),\ \exists x_2(\varphi(a_1, f(a_1)) = \varphi(a_2, x_2)),\\ \exists x_1 x_2(\varphi(a_1, x_1) = \varphi(a_2, x_2))\end{array}}{\begin{array}{c}\to \varphi(a_1, f(a_1)) = \varphi(a_2, g(a_1, a_2)),\ \exists x_2(\varphi(a_1, f(a_1)) = \varphi(a_2, x_2)),\\ \exists x_1 x_2(\varphi(a_1, x_1) = \varphi(a_2, x_2))\end{array}}\ (1a)}{\dfrac{\to \exists x_2(\varphi(a_1, f(a_1)) = \varphi(a_2, x_2)),\ \exists x_1 x_2(\varphi(a_1, x_1) = \varphi(a_2, x_2))}{\to \exists x_1 x_2(\varphi(a_1, x_1) = \varphi(a_2, x_2))} \to \exists_2} \to \exists_2$$

will be a deduction of this sequent in $L_2^=$.

In L_2^* the sequent $\to \exists x_1 x_2(\varphi(a_1, x_1) = \varphi(a_2, x_2))$ is not an axiom. The rule $\to \exists_1$ is the only rule with whose help this sequent can be obtained in L_2^*. Under this restriction to the rule $\to \exists_1$ it is permissible to substitute into the quasi-formula $\varphi(a_1, x_1) = \varphi(a_2, x_2)$, in place of the bound variables x_1 and x_2, the free variables a_1 or a_2 or terms beginning with the function symbol φ. For the sequent under consideration to be deducible in L_2^* it is necessary to substitute $f(a_1)$ into the quasi-formula $\varphi(a_1, x_1) = \varphi(a_2, x_2)$ in place of x_1 and $g(a_1, a_2)$ in place of x_2.

Consequently $\to \exists x_1 x_2(\varphi(a_1, x_1) = \varphi(a_2, x_2))$ is not deducible in L_2^*.

BIBLIOGRAPHY

1. G. Gentzen, *Untersuchungen über das logische Schliessen,* Math. Z. **39** (1934/35), 176–210, 405–431; Russian transl. in Mathematical Theory of Logical Inference, "Nauka", Moscow, 1967, pp. 9–74; English transl. in M. E. Szabo, *The collected papers of Gerhard Gentzen,* Studies in Logic and the Foundations of Math., North-Holland, Amsterdam, 1969, pp. 68–103. MR **36** #4957; **41** #6660.

2. S. Kanger, *Provability in logic,* Stockholm Studies in Philosophy. I, Almqvist and Wiksell, Stockholm, 1957. MR **19**, 239.

3. V. A. Matulis, *Two versions of classical computation of predicates without structural rules of deduction,* Dokl. Akad. Nauk SSSR **147** (1962), 1029–1031 = Soviet Math. Dokl. **3** (1962), 1770–1772. MR **26** #1241.

4. ———, *Variants of the classical predicate calculus with a single deduction tree,* Dokl. Akad. Nauk SSSR **148** (1963), 768–770 = Soviet Math. Dokl. **4** (1963), 176–178. MR **26** #4894.

5. H. B. Curry, *Foundations of mathematical logic,* McGraw-Hill, New York, 1963; Russian transl., Moscow, 1969. MR **26** #6036.

6. S. Kanger, *A simplified proof method for elementary logic,* Computer Programming and Formal Systems, North-Holland, Amsterdam, 1963, pp. 87–94; Russian transl. in Theory of Logical Inference, "Nauka", Moscow, 1967, pp. 200–209. MR **27** #1350.

7. M. G. Rogava, *The sequential versions of the applied predicate calculi,* Zap. Naučn. Sem. Leningrad. Otdel. Mat. Inst. Steklov. (LOMI) **4** (1967), 189–200 = Sem. Math. V. A. Steklov Math. Inst. Leningrad **4** (1967), 77–81. MR **39** #1293.

8. R. A. Pljuškevičus, *A sequential variant of constructive logic calculi for normal formulas not containing structural rules,* Trudy Mat. Inst. Steklov. **98** (1968), 155–202 = Proc. Steklov Inst. Math. **98** (1968), 175–229. MR **38** #4282; **43** #4620.

9. A. Ju. Pljuškevičene, *Elimination of cut type rules in axiomatic theories with equality,* Zap. Naučn. Sem. Leningrad. Otdel. Mat. Inst. Steklov. (LOMI) **16** (1969), 175–184 = Sem. Math. V. A. Steklov Math. Inst. Leningrad **16** (1969), 90–94. MR **42** #39.

10. G. E. Minc, *A variation of the tactics of the search for inference in sequential calculi,* Zap. Naučn. Sem. Leningrad. Otdel. Mat. Inst. Steklov. (LOMI) **4** (1967), 134–151 = Sem. Math. V. A. Steklov Math. Inst. Leningrad **4** (1967), 52–59. MR **41** #1505.

11. M. G. Rogava, *The elimination of cuts from some applied calculi,* Abstracts of Brief Scientific Communications, Internat. Congress Math., Moscow, 1966, Section I, p. 24. (Russian)

12. ————, *A sequential version of the calculus of equivalences,* Thbilis. Sahelmç. Univ. Gamoqeneb. Math. Inst. Šrom. **2** (1969), 9–11. (Russian) MR **43** #1812.

13. V. P. Orevkov, *On Glivenko sequent classes,* Trudy Mat. Inst. Steklov. **98** (1968), 131–154 = Proc. Steklov Inst. Math. **98** (1968), 147–176. MR **39** #2606; **43** #4620.

14. S. C. Kleene, *Introduction to metamathematics,* Van Nostrand, Princeton, N. J., 1952; Russian transl., IL, Moscow, 1957. MR **14**, 525; **19**, 2.

15. N. A. Šanin, G. V. Davydov, et al., *An algorithm for a machine scan of a natural logical deduction in a propositional calculus,* Academy of Sciences of the USSR, V. A. Steklov Mathematical Institute, Leningrad Department, "Nauka", Moscow, 1965. (Russian) MR **33** #5405.

16. S. C. Kleene, *Permutability of inferences in Gentzen's calculi LK and LJ. Two papers on the predicate calculus,* Mem. Amer. Math. Soc. No. 10 (1952); Russian transl. in Mathematical Theory of Logical Inference, "Nauka", Moscow, 1967, pp. 208–236. MR **14**, 439; **36** #4957.

17. V. A. Lifšic, *A normal form for conclusions in the predicate calculus with equality and functional symbols,* Zap. Naučn. Sem. Leningrad. Otdel. Mat. Inst. Steklov. (LOMI) **4** (1967), 58–64 = Sem. Math. V. A. Steklov Math. Inst. Leningrad **4** (1967), 21–23. MR **38** #5583.

18. ————, *Specialization of the form of deduction in the predicate calculus with equality and function symbols.* I, Trudy Mat. Inst. Steklov. **98** (1968), 5–25 = Proc. Steklov Inst. Math. **98** (1968), 1–24. MR **40** #5415; **43** #4620.

19. G. E. Minc, *Herbrand's theorem,* Math. Theory of Logical Deduction, "Nauka", Moscow, 1967, pp. 311–350. (Russian) MR **36** #4958.

20. A. Janiczak, *Undecidability of some simple formalized theories,* Fund. Math. **40** (1953), 131–139. MR **15**, 669; 1140.

21. M. Presburger, *Über die Vollständigkeit eines gewissen Systeme der Arithmetik ganzer Zahlen, in welchen die Addition als einzige Operation hervortritt,* C. R. du I Congrès Math. des Pays Slaves, Warsaw, 1930, pp. 92–101.

22. M. G. Rogava, *A specialization of certain logical rules in applied predicate calculi,* Sakharth. SSR Mecn. Akad. Moambe **56** (1969), 273–276. (Russian) MR **43** #1809.

Translated by F. M. GOLDWARE

ERRATA

TO THE CALCULI OF SYMBOLIC LOGIC. I

Proceedings of the Steklov Institute of Mathematics **98** (1968)

Page	Line	Printed	Should be
1	-6	$\mathfrak{c}_1, \mathfrak{c}_2$	$\mathfrak{x}_1, \mathfrak{x}_2$
1	-5	$\mathfrak{x}_1, \mathfrak{x}_2$	$\mathfrak{c}_1, \mathfrak{c}_2$
17	$-2, -1$	$\cdots s_2$, the fact that $\cdots$ implies that $s_1 \mathcal{O} s_2$.	$\cdots s_2$, such that $r_1 \mathcal{O} r_2$ and s_2 can be obtained from s_1 by replacing some occurrences of r_1 by r_2, for any term r of $s_1 \mathcal{O} s_2$, $r \mathcal{O} s_1$ implies $r \mathcal{O} s_2$.
21	18	t	t_1
22	3, 4	$\cdots$ implies that $t'_2 \mathcal{O} t_2$. But by $\cdots$ that $t_1 \mathcal{O} t_2$. Hence $\cdots$	$\cdots$ implies that $t_1 \mathcal{O} t_2$. Hence $\cdots$
22	$-11, -10$	R_1	R_2
75	-7	say	shall assume
75	-6	$\exists x_{i,1}, \cdots, \exists x_{i,k_i}$,	$\exists x_{i,1} \cdots \exists x_{i,k_i}$,
147	-13	(see [3])	(the sign of an occurrence of a logical symbol in a formula or sequent coincides with the sign of the occurrence (cf. [3] or [4]) of the domain of action of this occurrence of the logical symbol)
148	18	$\rightarrow \Delta''$	$\rightarrow \Delta'$
151	7, 8	$\cdots$ symbol" and "the $\cdots$ axioms" are defined $\cdots$	$\cdots$ symbol" are defined $\cdots$
151	9	$\cdots$ calculus G.	$\cdots$ calculus G. We shall say that an occurrence of a formula in a sequent is traceable to the axioms if this occurrence is a descendant of a principle occurrence of some axiom.
159	12	$\forall \rightarrow$	$\rightarrow \forall$
159 161	-12 6	C^+	C^*
164 165	-19 6, 7	above which there are no	for which there are above some premise no
169	6	*calculus,*	*calculus with equality,*